11+ Maths Practice Papers

BOOK 1-4

11+ MATHS ACADEMY

ISBN: 191254704X
ISBN-13: 978-1912547043

CONTENTS

INTRODUCTION

Key stage 2 practice papers have been provided in this book to allow your child to practice questions in their own time throughout the year. It will enable them to get a good understanding of their strengths and weaknesses. By covering all key topics, Eleven Plus Maths Academy helps your child to identify their weak areas and helps increase their confidence in tackling them during the real exam.

Answers and detailed mark schemes have been provided for you to assess how your child has performed. Eleven Plus Maths Academy also provide online support to help boost your child's scores.

When administering the tests you should provide your child with a quiet environment where they can complete the test under exam conditions and undisturbed.

There are 16 papers provided in this book. Each paper lasting 45 minutes each.

Please provide your child with the following equipment:

- ✓ a pen and pencil;
- ✓ a ruler (showing centimeters and millimeters);
- ✓ a protractor (angle measurer);
- ✓ a mirror; and
- ✓ tracing paper.

MARKING THE TESTS

The number in each box at the side of the page tells you the maximum number of marks for each question.
Use the answers and mark schemes provided at the back of this book to mark the papers.

Please note: these tests are only a guide as to the mark your child can achieve and cannot guarantee that the same mark is achieved during their Key Stage 2 tests.

Mathematics
Key stage 2

Paper 1

Calculator NOT allowed

For marker's use only

Page	Marks
Total	

Instructions

You **must not** use a calculator to answer any questions in this test.

Work as quickly and as carefully as you can.

You have **45 minutes** for this test.

If you cannot do one of the questions, **go on to the next one.**

You can come back to it later, if you have time.

If you finish before the end, **go back and check your work.**

Follow the instructions for each question carefully.

This shows where you need to put the answer.

If you need to do working out, you can use any space on a page.

Some questions have an answer box like this:

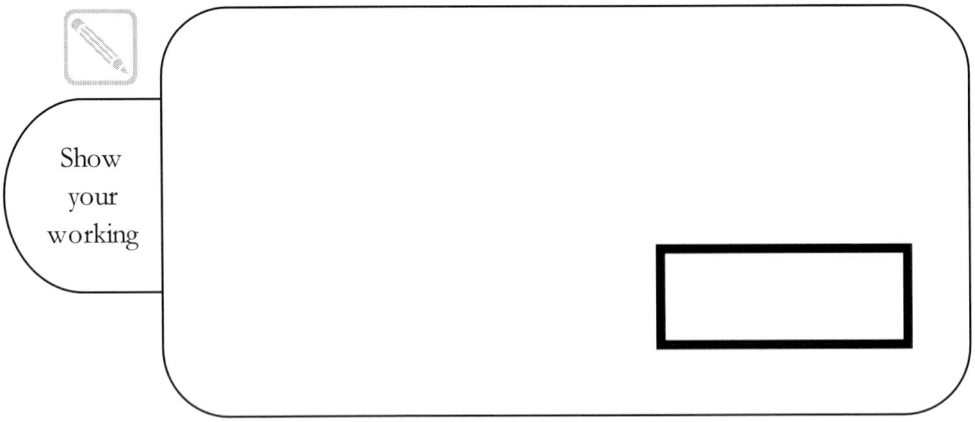

For these questions you may get a mark for showing your working.

1. Calculate.

14896 - 5348 =

56 x 39 =

2 Marks

2. What is the 5/8 of 128?

What is the 1 2/8 of 128?

2 Marks

3. Divide 2896 by 8.

1 Mark

4. Multiply 58 and 96

1 Mark

5. Please find the perimeter of the shape below.

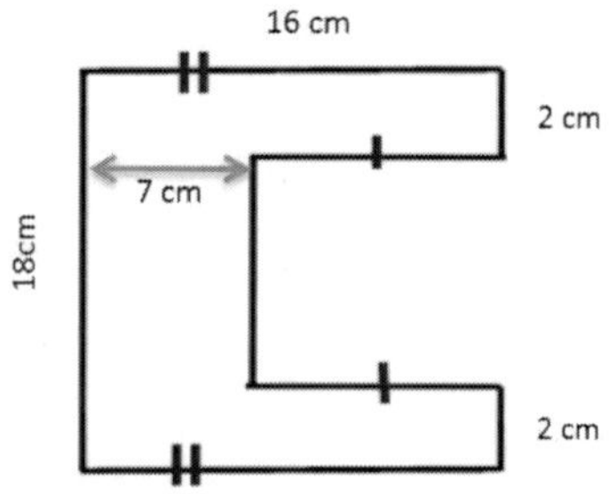

Show your working

2 Marks

6. Following are the prices of groceries in a store.

Item	Price
Apple	78p / 100grams
Milk	88p / 1 litre
Cake	£1.5/ 500grams
Chocolate	90p/ 200grams

Sara buys the following.

- Apple 500 grams
- Milk 500 millilitre
- Cake 750 grams
- Chocolate 500 grams

What is the total cost of the items Sara purchased?

2 Marks

If she gives a £20 note, how much balance will she get?

1 Mark

7. Which number in between 50 and 60 can be divided exactly by both 7 and 8?

1 Mark

8. Below are the details of three children.

Jack	Peter	Joanne
32.75Kg	**38.23Kg**	**29.98Kg**

What is the total weight of all three children?

1 Mark

How much heavier is Peter than Joanne?

1 Mark

If Jack reduces 1kg, what is the mean weight of all three children?

1 Mark

9. Calculate.

$X = 6$

$X2 + (3 + 5x) – 4x =$

Show your working

2 Marks

10. Add the below numbers together.

- Two hundred and sixty eight
- Three thousand and ninety two
- Nine thousand six hundred and six

Show your working

1 Mark

11. Fill the gaps.

9586 - ☐ = 2957

☐ - 3986 = 1432

2 Marks

12. Amy is 9 years and 3 months old. Her brother John is 2 years and 7 months older to Amy. How old is John?

Show your working

☐

1 Mark

13. Last year a school bought 189 new chairs and removed 53 old chairs. A charity donated them 236 chairs. At the end of the year they had 1985 chairs.

How many chairs did they have in the beginning of last year?

Show your working

2 Marks

14. Joanne has only 2 pence and 5 pence coins in her till. Altogether she has 30 coins. She counts all the coins and says that she has £1.20. She also says the number of 5 pence is double the amount of 2 pence. How many 2 pence and 5 pence she had in the till?

Show your working

2 Marks

15. In a bag there 12 green balls, 5 red balls and 3 blue balls.

- If John closes his eyes and picks one ball, what is the probability of getting a red ball?

- Sara takes out two green balls. Now what is the probability of John getting a red ball?

2 Marks

16. Calculate.

15 x (28 -4) ÷ 2 + 25 =

2 Marks

17. Sara gets 17 questions right in a test out of 40 questions. What percentage of the questions did she get right?

2 Marks

18. If a square's perimeter is 64cm, what is the area?

1 Mark

19. If the time is 9.17p.m, what is the time in 24 hour clock?

1 Mark

20. Three elephants weigh 1 ton 500grams. What is the total weight in grams?

1 Mark

21. What is the next two numbers?

1, 5, 9, 13, _, _

,

2 Marks

22. Find b.

12a + 15b = 294

6a – 2 = 70

Show your working

2 Marks

23. What is the difference between 1.25 metre and 278 millimetres?

1 Mark

24. How many eights are there in 1264?

1 Mark

[END OF TEST]

Please do not write on this page.

Mathematics
Key stage 2

Paper 2

Calculator NOT allowed

For marker's use only

Page	Marks
Total	

Instructions

You **must not** use a calculator to answer any questions in this test.

Work as quickly and as carefully as you can.

You have **45 minutes** for this test.

If you cannot do one of the questions, **go on to the next one.**

You can come back to it later, if you have time.

If you finish before the end, **go back and check your work.**

Follow the instructions for each question carefully.

This shows where you need to put the answer.

If you need to do working out, you can use any space on a page.

Some questions have an answer box like this:

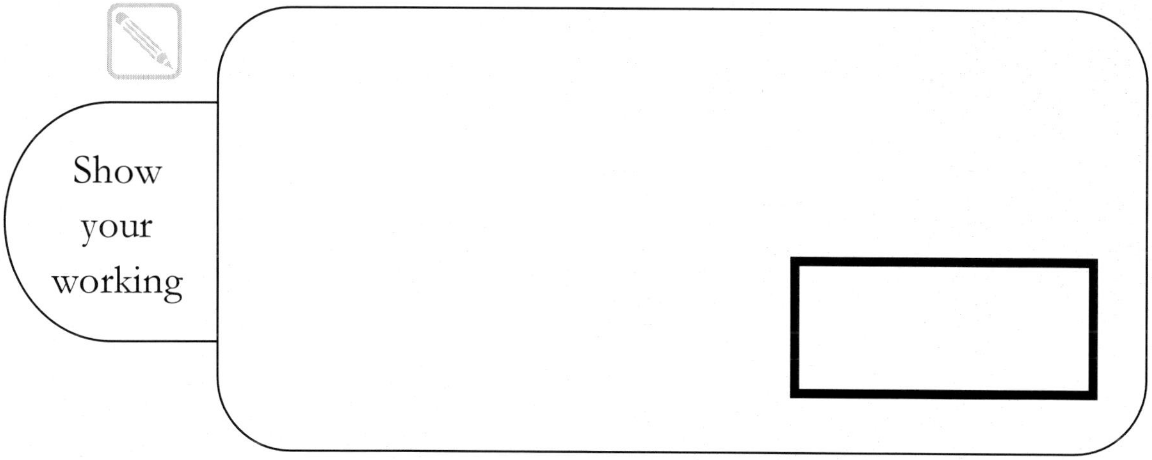

For these questions you may get a mark for showing your working.

1. Put the following in order of size, starting from the smallest.

¼ 2/5 6/8 ½

1 Mark

2. If two numbers multiply by 35 and their total is 12 when adding, what are the two numbers?

2 Marks

3. Amy takes 36 minutes to reach her school. What time does she have to leave home if she has to reach school at quarter past eight?

1 Mark

4. Fill in the blanks.

Example - The first 5 multiples of 3 add ups to 45
(3+6+9+12+15= 45)

The first 3 multiples of ______ add ups to 42

The first ______ multiples of 4 add ups to 60

2 Marks

5. Shade 2/3 of this pattern

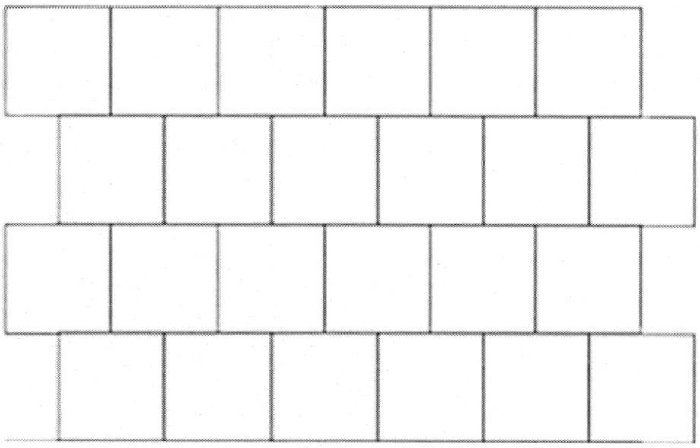

2 Marks

6. The table shows the weight of four children

Name	Weight
Tina	28kg
Peter	34kg
Ben	38kg
Rita	36kg

Amy says the average weight of all four children is equal to Peter's weight.

Is Amy correct?

Explain why?

2 Marks

7. Jacob stays in a hotel for 9 days. He pays £530; this includes the hotel charges for 9 days and an additional charge for a bottle of water which cost £8. How much does that hotel charge for one day?

Show your working

2 Marks

8. Rita played in the park from 5.15pm to 7.10 pm. How long did she spend in the park?

1 Mark

9. The diagram shows the favourite fruit of 90 children in a class.

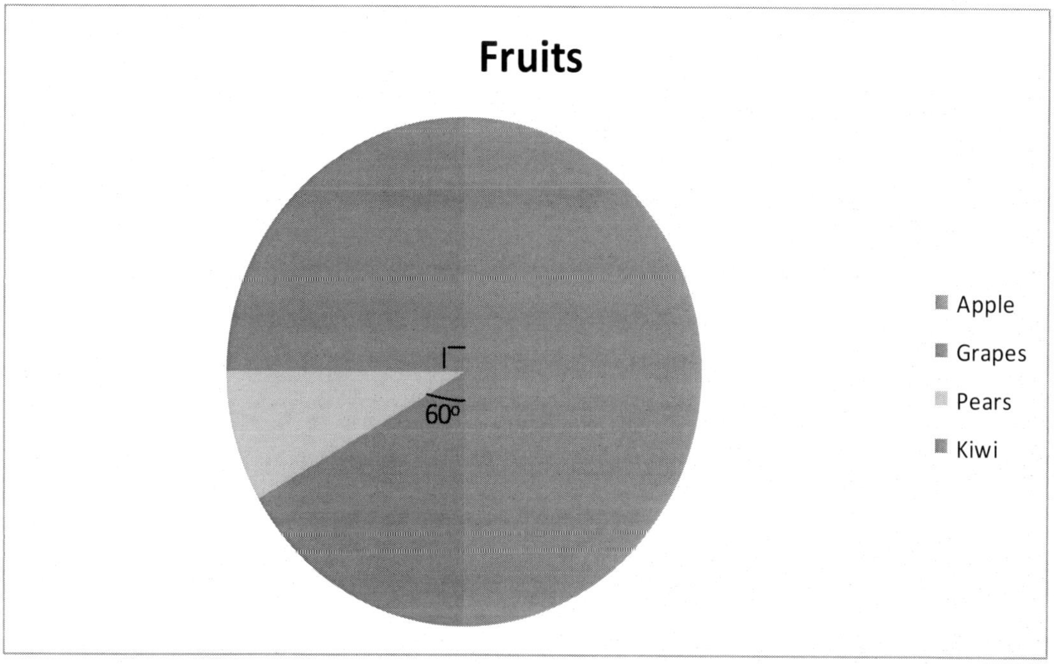

- What angle has been used to indicate pears?

1 Mark

- How many children in the class like grapes?

1 Mark

10. The second hand of the clock points at number 2. After 90 seconds what number will it be pointing to?

1 Mark

11. The time in USA is 5 hours behind UK. Rita wants to be the first to call her friend in USA for her birthday at USA time 12 am. At what time does she have to call her at the UK time?

2 Marks

12. There are few number cards as shown below.

9	5	8	3	2

Using all these cards once only
Make the largest number

Make the smallest odd number

2 Marks

13. Amy had marbles in a bag.

To play a game with 5 groups, she gave 7 marbles to each group of children. Therefore, one marble was left off in the bag.

For another game with 7 groups, she gave 5 marbles to each group and one marble was left off.

For the final game with 6 groups, she gave 6 marbles for each group and had no marble left.

How many marbles did she have in the bag?

Show your working

2 Marks

14. Two numbers in a ratio 2:5.

One number is 30.

There are two possible values for the other number. What are the two possible values?

2 Marks

15. In a class room, children were asked what their favourite ice cream was. The result is shown in the below Venn diagram.

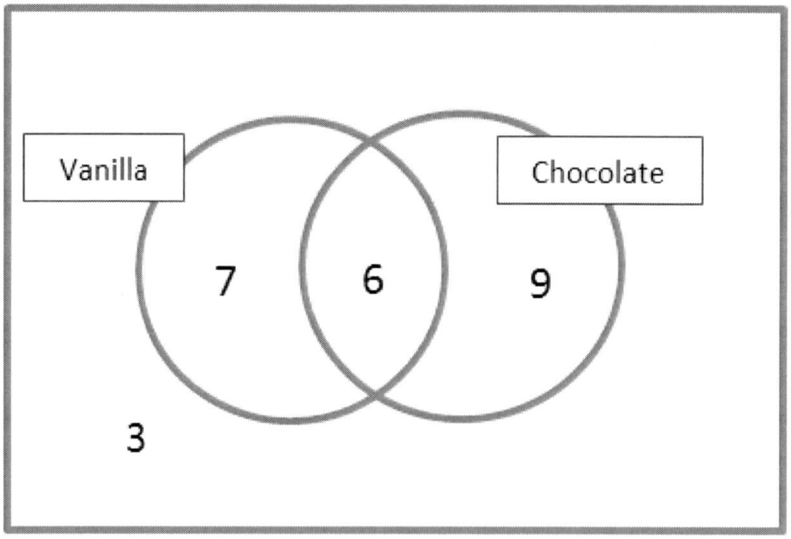

What percentage of the class room like both vanilla and chocolate ice cream?

1 Mark

What percentage of the class room does not like both vanilla and chocolate ice cream?

1 Mark

What percentage of the class room does not like vanilla ice cream?

1 Mark

16. How many seconds a baby has lived when he/she is one year old?

1 Mark

17. There is a cuboid tank full of water. The measurement of the tank is as below.

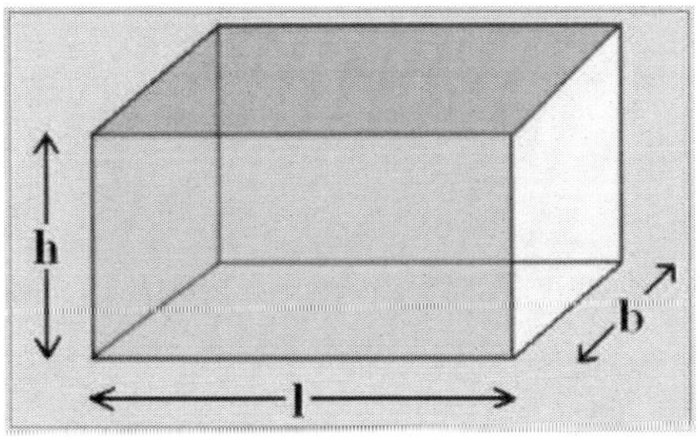

h	10cm
l	20cm
b	15cm

Amy tries to remove all the water from the tank by using a bucket. The bucket can contain 150cm3 of water at a time. If she fills the bucket fully, how many times does she have to use the bucket to clear all the water in the tank?

Show your working

2 Marks

18. A box of chocolates cost £65. There are 25 chocolate bars in the box. Each bar contains 5 pieces of chocolate. How much does a piece of chocolate cost?

1 Mark

19. Find b.

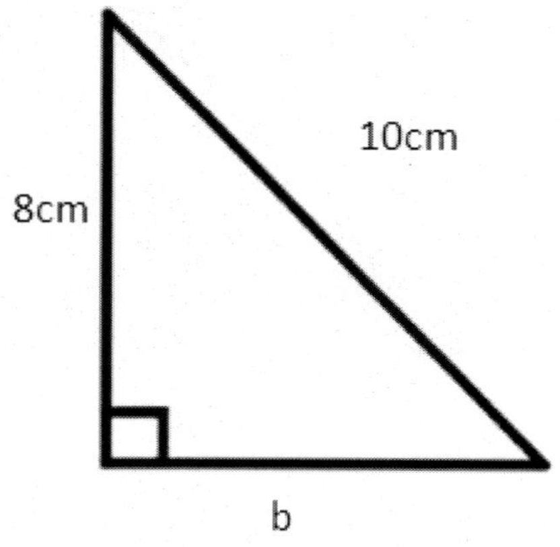

2 Marks

20. What fraction of an hour is 20 minutes?

1 Mark

21. A bottle can contain 5.25 litres. It is 70% filled with water. How many more water is need to fill the bottle fully? Provide the answer in millilitre.

1 Mark

22. 8 boxes of chocolate cost £10. How much does 10 boxes of chocolate cost?

1 Mark

23. Find B.

$$\begin{array}{r} 6\ B \\ \times \quad B \\ \hline 6\,2\,1 \end{array}$$

2 Marks

24. John prints his project report. The total cost for it is £5.65. Below are the prices for printing at the store.

Printing (per page)	25p
Binding	£1.05
Cover	85p

How many pages did John print?

Show your working

2 Marks

[END OF TEST]

Please do not write on this page.

Mathematics
Key stage 2

Paper 3

Calculator NOT allowed

For marker's use only

Page	Marks
Total	

Instructions

You **must not** use a calculator to answer any questions in this test.

Work as quickly and as carefully as you can.

You have **45 minutes** for this test.

If you cannot do one of the questions, **go on to the next one.**

You can come back to it later, if you have time.

If you finish before the end, **go back and check your work.**

Follow the instructions for each question carefully.

This shows where you need to put the answer.

If you need to do working out, you can use any space on a page.

Some questions have an answer box like this:

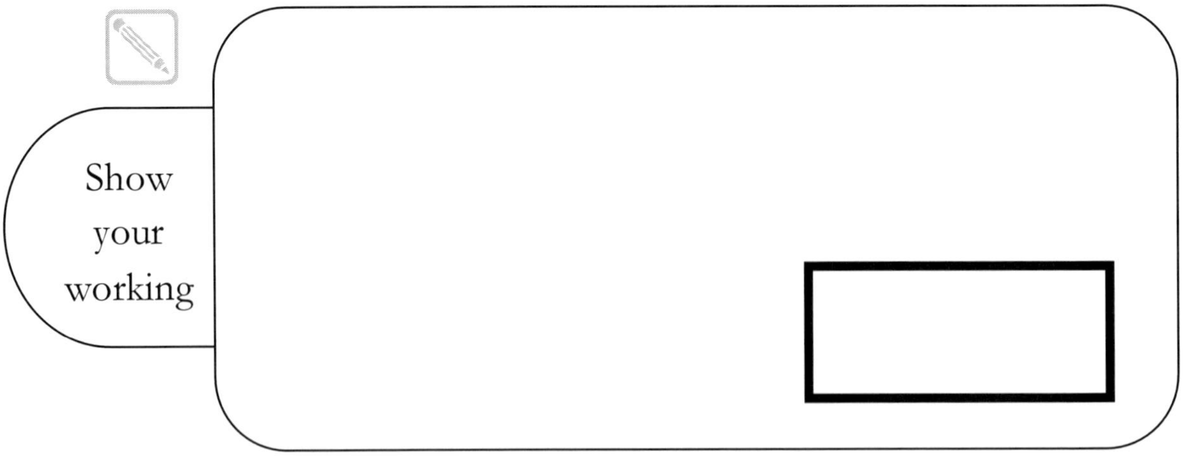

For these questions you may get a mark for showing your working.

1. Choose a number from each set to get the answer.

Set 1	Set 2	Set 3
4, 8, 5	2, 4, 3	3, 4, 10

☐ x ☐ - ☐ = 10

2 Marks

2. Lucy, Aaron and Sharon went to the garden to pluck fruits.
The below table shows the number of fruits they collected.

Apple					
Orange					
Strawberry					

Key	
	Represents 10 apples
	Represents 10 oranges
	Represents 10 strawberries

How many apples have they plucked more than oranges?

1 Mark

All together how many fruits have they plucked?

1 Mark

If they sell each strawberry for 40p, how much will they gain?

1 Mark

3. Aaron buys a book for £1.60. He gives three coins to the seller and gets 40p as balance.
What are the three coins he would have given?

Show your working

1 Mark

4. Sharon makes a 3D shape using 2 triangles and 3 rectangles.
What shape is Sharon making?

1 Mark

Draw it.

1 Mark

5. Calculate.

$190 - (11 - 8)^2 \times 12 =$

2 Marks

6. Aaron has x number of candies in a box. He equally divides it among 12 friends. Each person gets 9 candies. How many candies were there in the box?

Show your working

1 Mark

7. Sharon goes to school in her bicycle. She rides in a speed of 5km/hour. Her school is 2km from her home. If she leaves home at 7.45am at what time will she arrive at school?

2 Marks

8. Fill in the blanks.

8	+	___	=	17
9	___	6	=	1.5
___	x	7	=	56

2 Marks

9. Calculate

568 x 1.6 =

1 Mark

10. Complete the number sequence

10, 25, 40, ___ , ___

38, 30, 22, ___ , ___

2 Marks

11. Draw the mirror image.

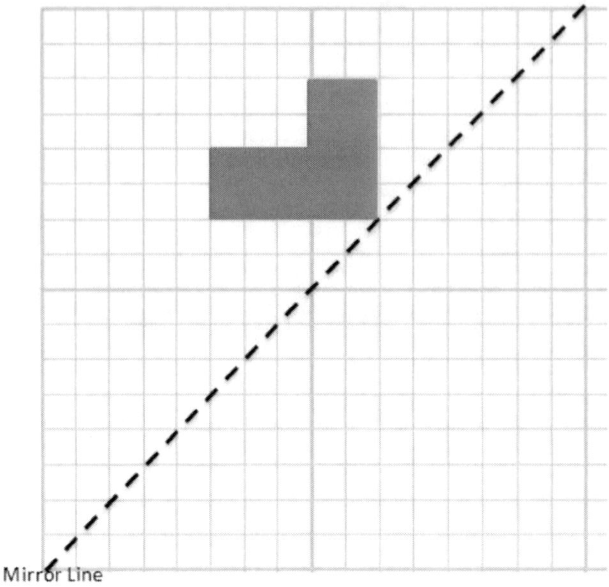

2 Marks

12. Calculate.

9/15 ÷ 27/35

1 Mark

13. Write the correct sign =, < and > in the blank grey boxes.

14	-	7		18	-	9
12	x	5		180	÷	3
15	+	8		19	+	3
30	-	16		14	+	9

2 Marks

14. Lucy and Sharon each buy a meal.

Lucy pays £10 and gets a balance of £1.25. Sharon's meal cost £3.65 more than Lucy's meal.

What is the price of Sharon's meal?

Show your working

1 Mark

15. What is the volume of the below object?

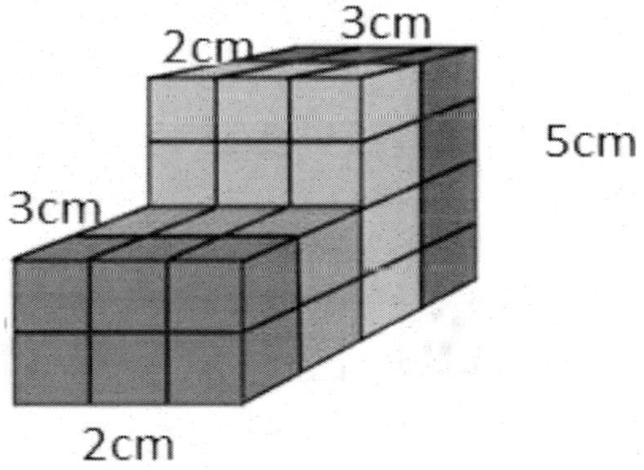

2 Marks

16. 10 apples weighs 1.5kg. Aaron buys 8 apples. What is the weight of the apples Aaron bought?

If 1kg apple cost £2.50, how much will Aaron pay for 8 apples?

Show your working

1 Mark

17. Calculate.

19586 ÷ 8 =

1 Mark

152 x 12 =

1 Mark

18. Add 0.182 1.82 + 18.2+ 182 =

1 Mark

19. Below is a sketch of a shape (not to scale).

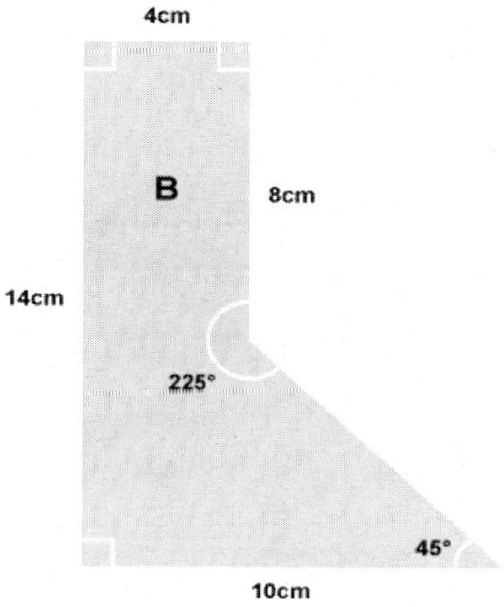

Use angle measurer and ruler and draw the shape accurately.

2 Marks

20. Lucy has a standard dice. When she spins it, what is the probability of getting an odd number?

1 Mark

21. Write the prices in order, starting from the smallest.

£6.96	£6.89	£0.99	£6.99

1 Mark

What is the difference between the highest price and the lowest price?

1 Mark

22. Calculate.

398 – 259.78 =

1 Mark

23. Sharon bought two apples and two oranges. Apple cost 20p more than orange.

The total cost was £1.20. What is the price of one apple and one orange?

Show your working

2 Marks

24. Three different numbers add up to 30. All three numbers are even and less than 15. What could the three different numbers be?

2 Marks

[END OF TEST]

Please do not write on this page.

Mathematics
Key stage 2

Paper 4

Calculator NOT allowed

For marker's use only

Page	Marks
Total	

Instructions

You **must not** use a calculator to answer any questions in this test.

Work as quickly and as carefully as you can.

You have **45 minutes** for this test.

If you cannot do one of the questions, **go on to the next one.**

You can come back to it later, if you have time.

If you finish before the end, **go back and check your work.**

Follow the instructions for each question carefully.

This shows where you need to put the answer.

If you need to do working out, you can use any space on a page.

Some questions have an answer box like this:

For these questions you may get a mark for showing your working.

1. Join the following equations to the nearest number

549 – 280	100
155 x 3	400
168 + 192	300
550 ÷ 5	500

2 Marks

2. Write the missing digits.

5 ☐ + ☐ 5 = 150

1 Mark

3.

Find x and y.

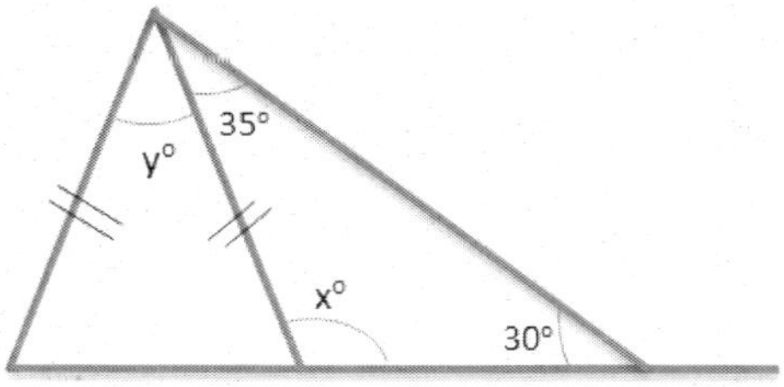

X - ☐

Y – ☐

2 Marks

4. The time is 11.25am. Lucy says the time is closer to 11am than 12pm.

Explain why?

2 Marks

5. Cross all the diagrams which are ¼ shaded.

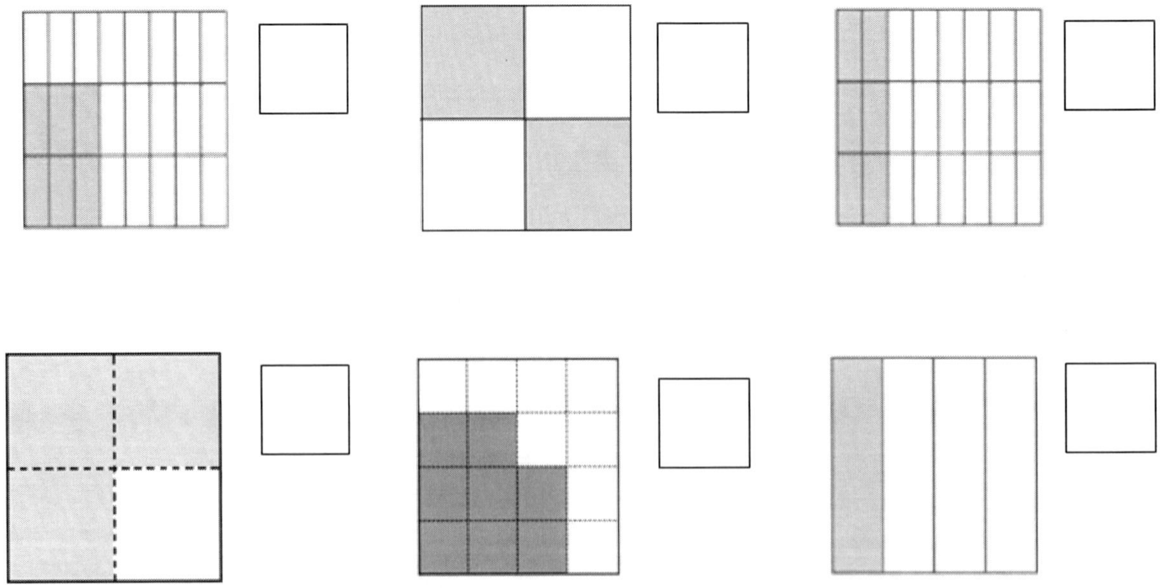

2 Marks

6. There are 40 children in a class. Each child needs 3 strawberries each. There are 20 strawberries in a box. How many boxes of strawberries do they need for all 40 children?

Show your working

2 Marks

7. Which is not a quadrilateral among the shapes below? Please circle.

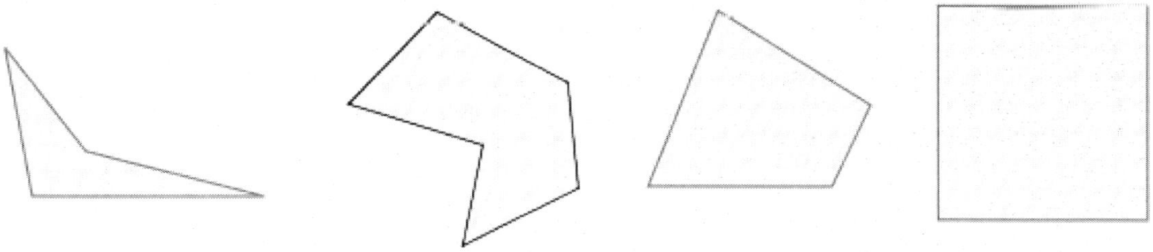

1 Mark

8. Write the below numbers in an order, starting from the smallest.

1.86 8.16 18.6 0.186

1 Mark

9. In a class test, Aaron finished in 18 minutes. Sharon finishes 5 minutes before Aaron. Lucy finishes 9 minutes after Sharon.

How long does Lucy take to finish the test?

1 Mark

If the test has started at 10.15am, at what time did Sharon finish?

1 Mark

10. Sharon buys a 250ml bottle of water. She drinks 70% of the water. How much water did Sharon drink? Provide your answer in ml.

1 Mark

11. Sharon and Lucy each have a till. Altogether they have £25. Sharon puts £2.50 to Lucy's till, so that they both have same amount of money.

How much money did each have in the start?

Sharon –

Lucy –

2 Marks

12. There is 30 marbles in a bag. Aaron takes 15 marbles. From the balance Sharon takes 60% of the marbles.

Who has the highest number of marbles?

1 Mark

Explain why?

1 Mark

13. Calculate.

2.8 + (5.6 – 2.8) + 3.9 =

1 Mark

14. Sharon rotates the arrow 270° anticlockwise about the centre of the grid. Draw the arrow in its new position.

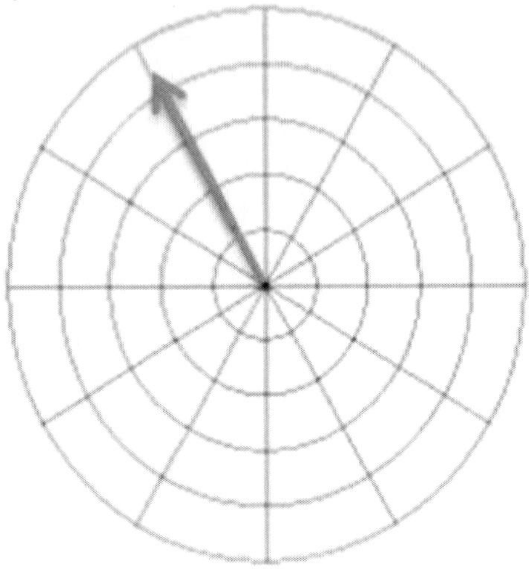

2 Marks

15. 1kg cake cost £2.80. 7 friends buy 9kg of cake and share the cost. How much does each person pay?

Show your working

2 Marks

16. Sharon has some rectangles and squares.

Using those she makes the shape below.

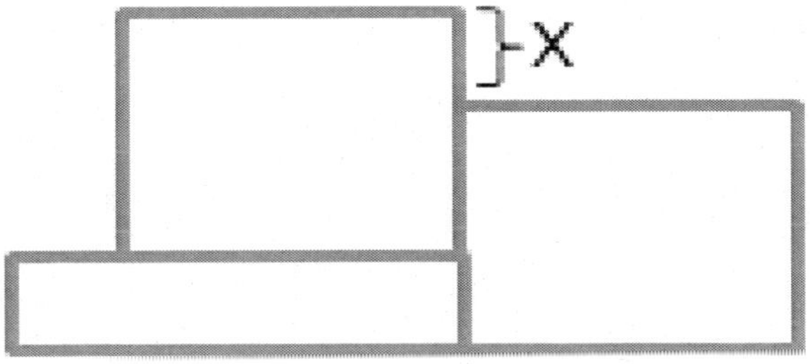

Find the length of X?

1 Mark

What is the area of the above image?

1 Mark

17. Two whole numbers between 30 and 50 make 1710 when multiplied. What are the two numbers?

,

1 Mark

18. Draw ten to seven in the below clock.

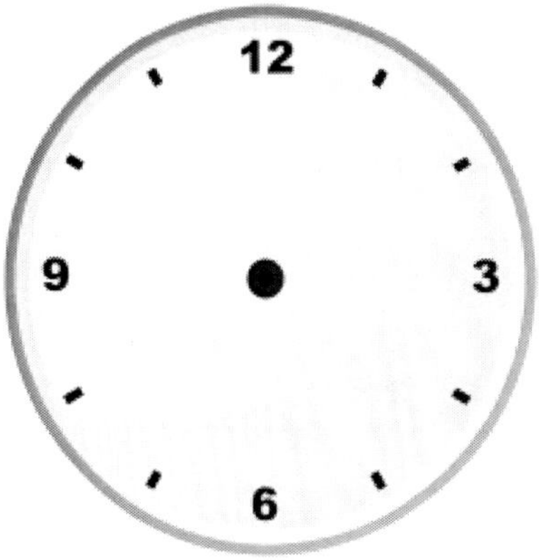

1 Mark

19. Aaron says "multiples of 10 ends in 0".

Is Aaron correct?

1 Mark

Explain why?

1 Mark

20. Write the missing number.

15 x ☐ x 9 = 1620

26 + ☐ - 16 = 29

2 Marks

21. Lucy buys 8 Roses and 5 Daffodils.
Roses cost £1.5 more than daffodils.
Altogether it cost £31.5.

Write an expression for this.

☐

1 Mark

How much does a single rose and daffodil cost?

Rose – ☐

Daffodil – ☐

1 Mark

22. Sharon, Aaron and Lucy go to a super market.

Their bills totals are as follow.

Sharon - £29.65

Lucy - £35.25

Aaron - £40.65

Aaron rounds each bill total to full number and adds it. What is the total amount Aaron gets?

1 Mark

Lucy adds up three bills exactly. What is the difference between Lucy's total and Aaron's total?

1 Mark

23. Write the smallest whole number to make the statement true

95 + ☐ > 153

1 Mark

24. Every 200g cake has 120g protein. A box of cake weighs 1kg and has 30 equal pieces. How much protein is there in on cake piece?

Show your working

2 Marks

[END OF TEST]

Please do not write on this page.

Mathematics
Key stage 2

Paper 5

Calculator NOT allowed

For marker's use only

Page	Marks
Total	

Instructions

You **must not** use a calculator to answer any questions in this test.

Work as quickly and as carefully as you can.

You have **45 minutes** for this test.

If you cannot do one of the questions, **go on to the next one.**

You can come back to it later, if you have time.

If you finish before the end, **go back and check your work.**

Follow the instructions for each question carefully.

This shows where you need to put the answer.

If you need to do working out, you can use any space on a page.

Some questions have an answer box like this:

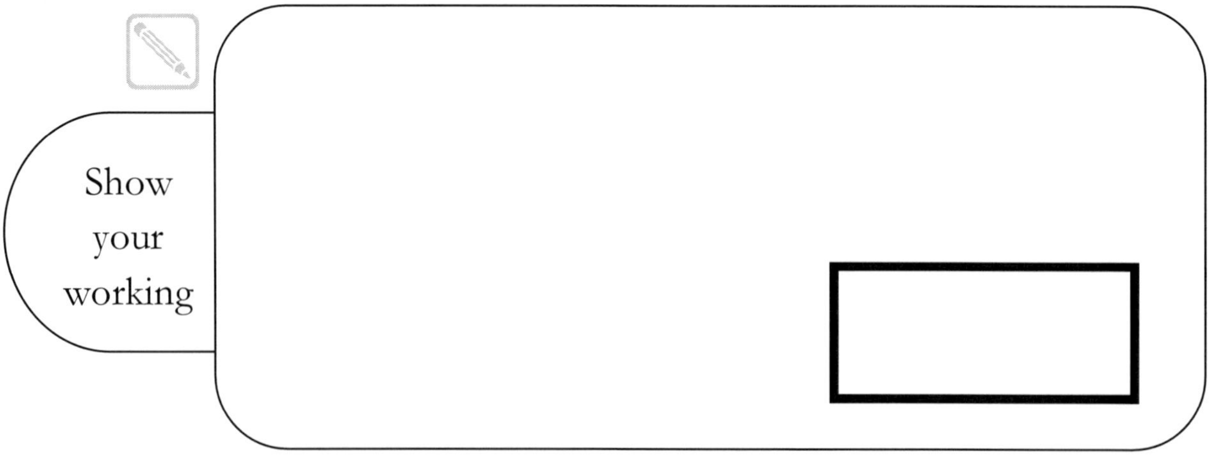

For these questions you may get a mark for showing your working.

1. Fill up the blanks.

11 ____ 35 47 59 ____

____ & ____

2 Marks

2. The numbers in the two circles add up to the number in square.

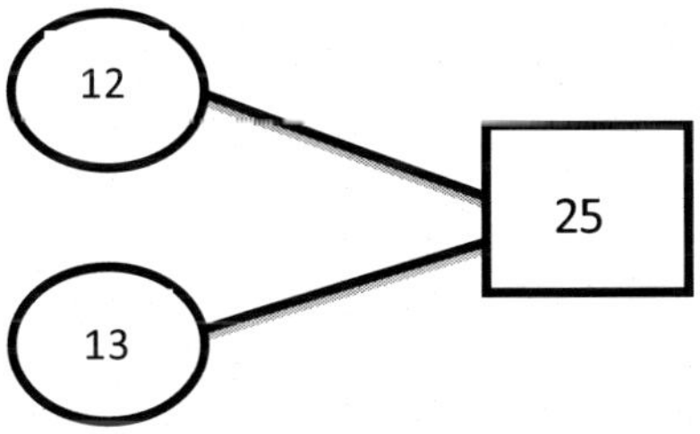

Using the same rule, write the missing numbers.

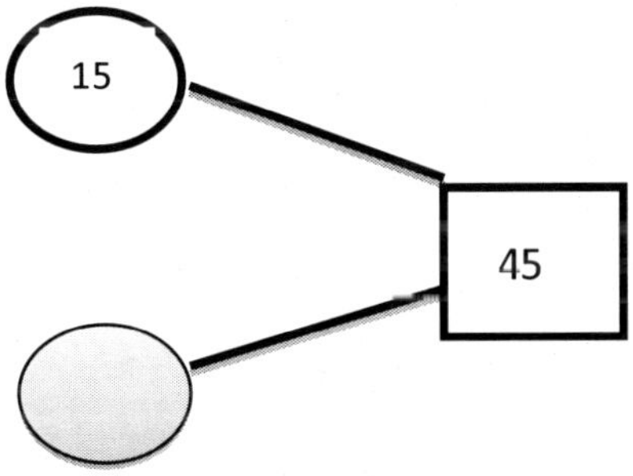

____

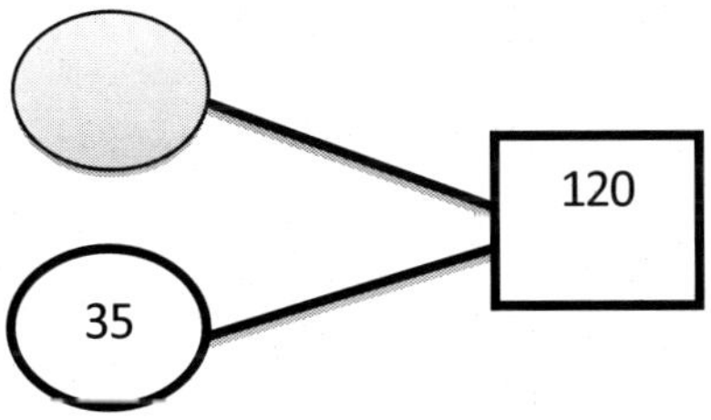

____

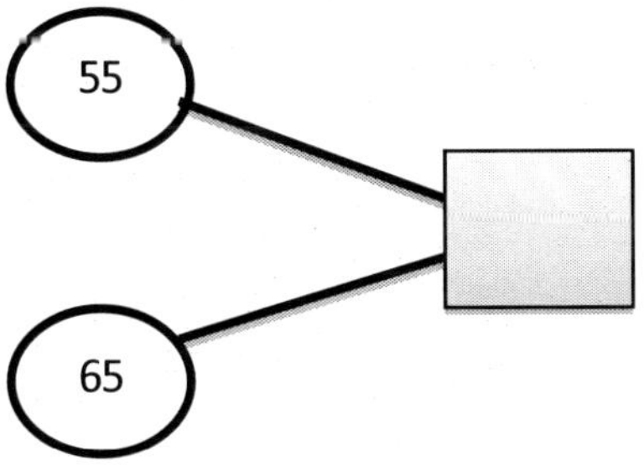

____

3 Marks

3. Circle the closest to 200.

228   186   2006   215   130

1 Mark

4. The number 6 goes in two of the squares of this multiplication grid.

Cross (X) the two squares where 6 goes.

×	1	2	3	4	5
1					
2					
3					
4					
5					

1 Mark

5. Complete these calculations.

20 x 10 = ☐

20 x ☐ = 2000

☐ ÷ 200 = 10

200 ÷ 20 = ☐

2 Marks

6. Calculate

$(12 \times 8) \div 4 + 18 =$

1 Mark

7. Amy buys two pens, each at the same price.
She pays £5 and gets £1.60 balance.
What is the cost of one pen?

1 Mark

8. Megan and Ben each have a bag.
Megan's bag weighs 2 ½ Kg.
Ben's bag weighs 2.3 Kg.
How many grams less does Ben's bag weigh than Megan's bag?

1 Mark

9. Chen has a parcel.
20% of the parcel weighs 54 Kg.
What is the weight of the 50% of the parcel?

1 Mark

10. Write these in the order of size, starting from the biggest.

2/4   0.35   3/5   0.65

1 Mark

11. What is the perimeter of the given shape?

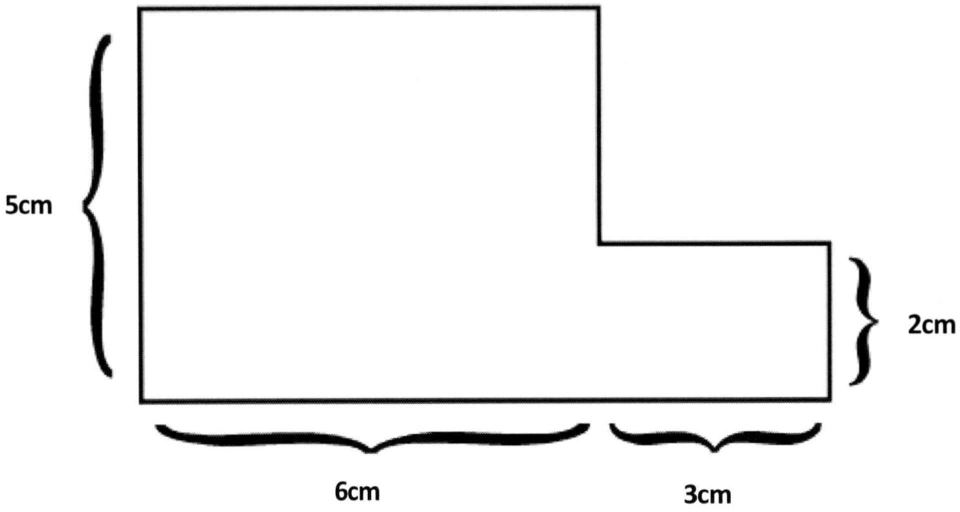

cm

1 Mark

12. Peter bought a pen, a pencil and a ruler for 85p.

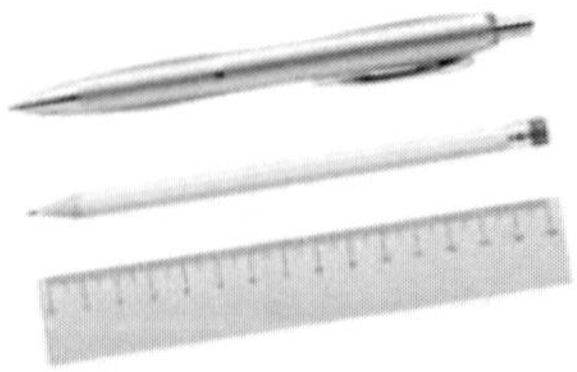

The pen cost 15p more than the pencil.
The ruler cost 5p less than the pencil.

How much does a pen, a pencil and a ruler each cost?

Show your working

Pen :

Pencil :

Ruler :

3 Marks

13. What are the sizes of angels b and e?

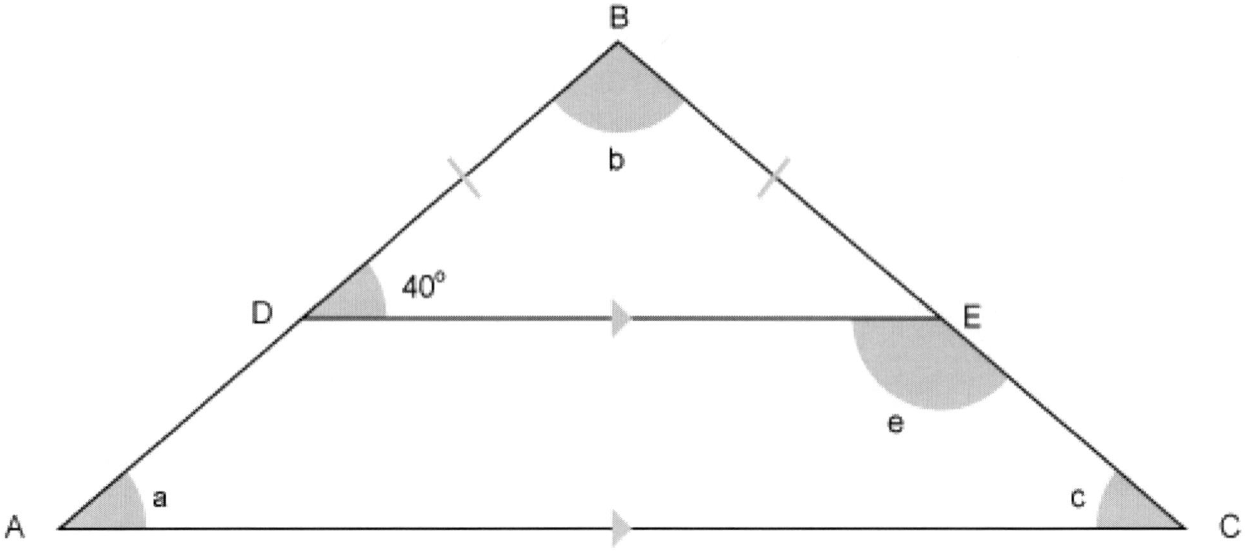

Angle b = ☐ Angle e = ☐

2 Marks

14. Sarah has a set of cards.

15 23 34 45

She chooses two of these cards and adds the number.
She rounds the result to the nearest 10.
Her answer is 70.
Which two cards did Sarah choose?

&

1 Mark

15. Write these times in order, starting with the shortest.

15 weeks	3 days	52 hours	3 months

1 Mark

16. Write the four missing numbers A, B, C and D.

Note – to make the number in the square, multiply the two numbers in the connecting circles.

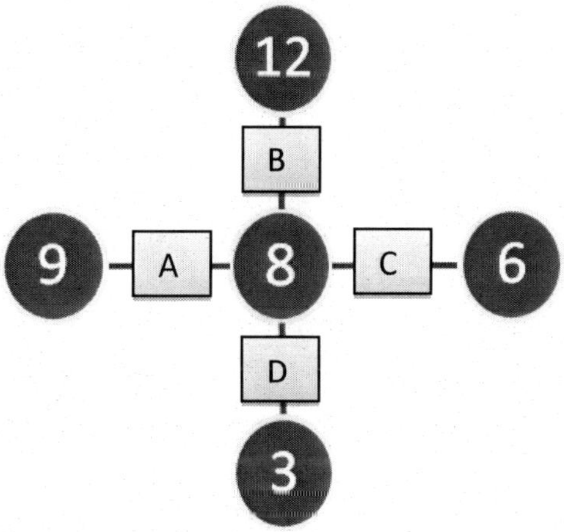

A =

C =

B =

D =

2 Marks

17. Draw the reflection of the shape in the mirror line.

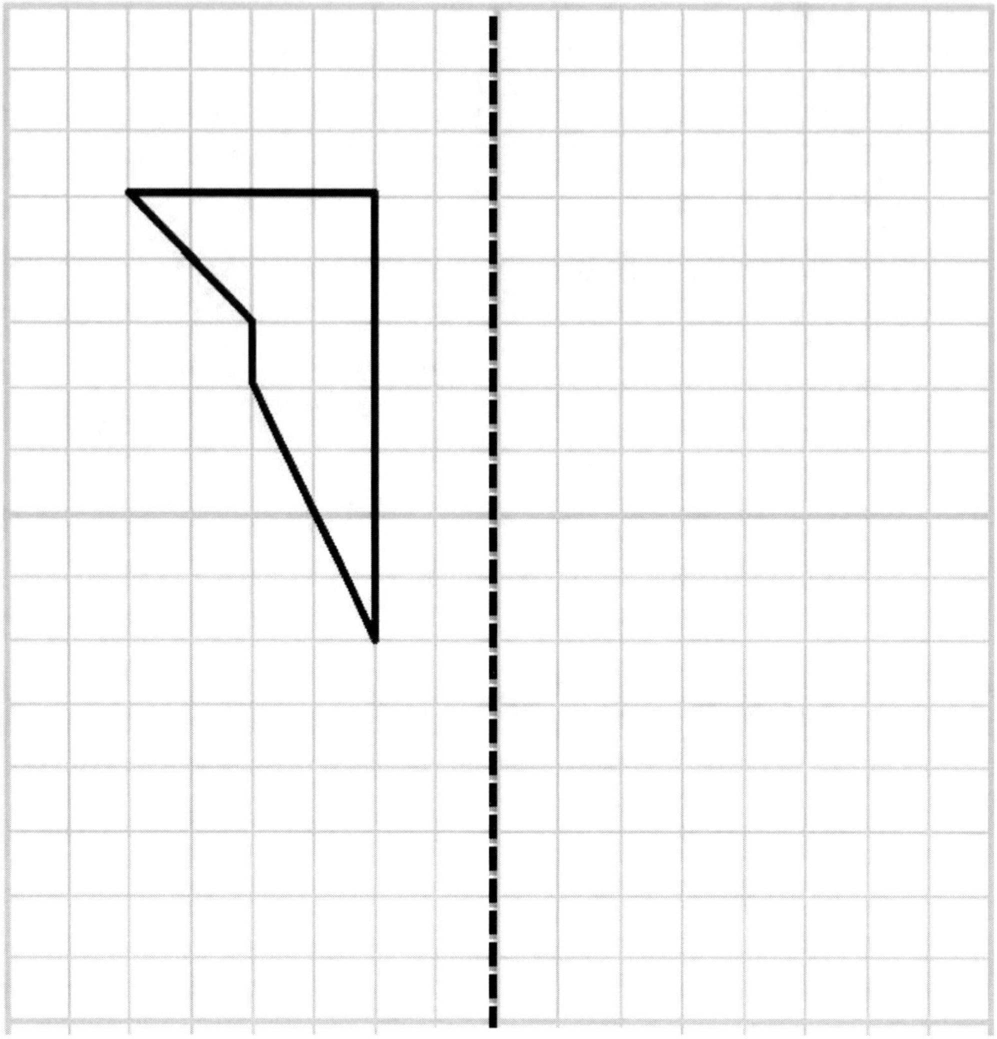

Mirror line

2 Marks

18. Provide the answer of the calculation.

$9 + (5 \times 2) \div 5 =$

1 Mark

19. Martha buys a meal which includes the following.

Item	Cost
Burger	£4.15
French fries	£2.50
Coke	£1.95

What is the cost of the meal?

Show your working

£

2 Marks

20. Calculate.

$925 \div 16 =$

1 Mark

21. The following table shows all the possible totals by adding all the possible combinations from Dice A and Dice B.

Dice A (columns), Dice B (rows)

	1	2	3	4	5	6
1	2	3	4	5	6	7
2	3	4	5	6	7	8
3	4	5	6	7	8	9
4	5	6	7	8	9	10
5	6	7	8	9	10	11
6	7	8	9	10	11	12

What is the probability of throwing a total 6?

Provide the answer in percentage.

Show your working

%

2 Marks

22. The weight of potato sack is 11.25Kg.

What is the weight of the potato sack in grams?

1 Mark

23. $6y + 2 = 5y + 8$

What is the value of y?

1 Mark

24. Calculate.

$34.12 \div 1.2 =$

1 Mark

[END OF TEST]

Please do not write on this page.

Mathematics
Key stage 2

Paper 6

Calculator NOT allowed

For marker's use only

Page	Marks
Total	

Instructions

You **must not** use a calculator to answer any questions in this test.

Work as quickly and as carefully as you can.

You have **45 minutes** for this test.

If you cannot do one of the questions, **go on to the next one.**

You can come back to it later, if you have time.

If you finish before the end, **go back and check your work.**

Follow the instructions for each question carefully.

This shows where you need to put the answer.

If you need to do working out, you can use any space on a page.

Some questions have an answer box like this:

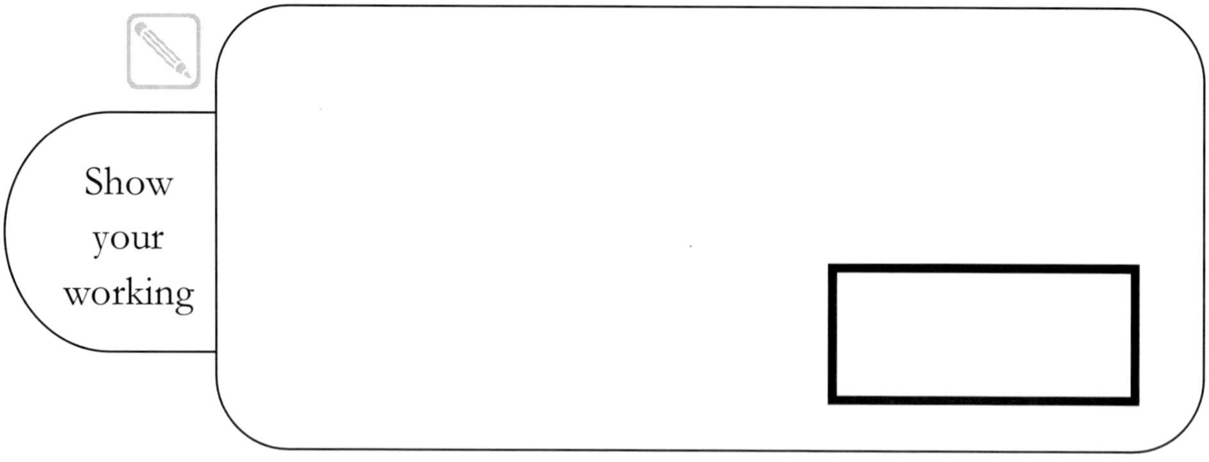

For these questions you may get a mark for showing your working.

1. Christy has two boxes

There are X counters of chocolates in each box.

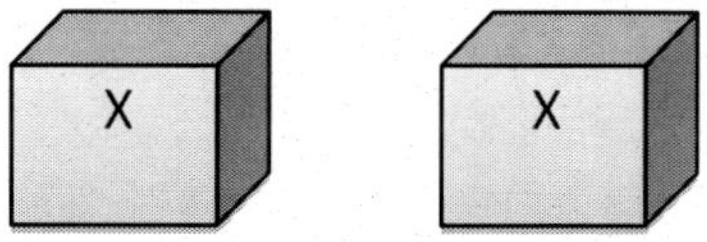

She puts all her chocolates in a bag, and then gives 3 chocolates to Seb.

Write an expression from the number of chocolates in the bag now.

1 Mark

2. Bill and Tom are brothers. Bill weighs two times more than Tom.

Bill Tom

Together they weigh 96Kg.

How much does Bill weigh?

2 Marks

3. Here is a triangle on a square grid. Draw the new position when it is rotated. 270° about point **X**.

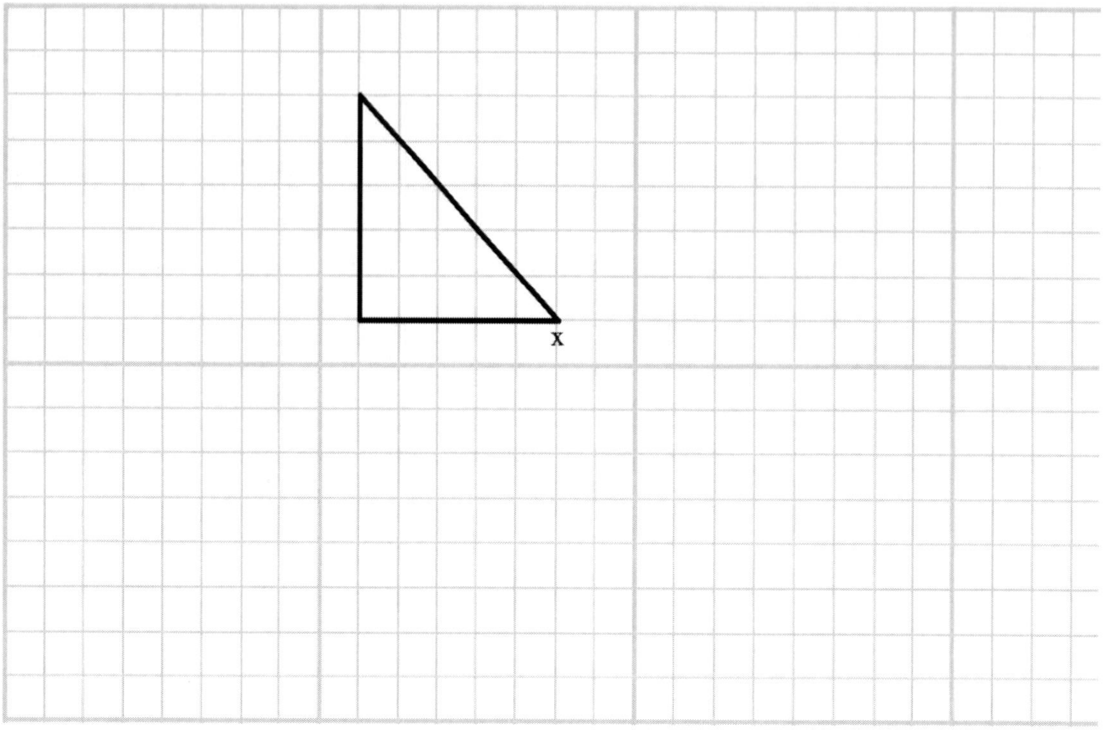

2 Marks

4. Amy has £120 to buy cupcakes and chocolates for her birthday.

Each cupcake costs £6.75
Each chocolate costs £3.75

She buys 15 cupcakes.
How many chocolates can she buy for the balance of the money?

1 Mark

5. The below table shows the number of gifts 4 children got for Christmas.

Name	Shoes	Book	Soft toy	Crayons
Nicky	1	2	1	2
Christy	2	1	2	2
Abby	1	2	2	1
Sara	2	1	1	1

Here is the graph information of three children.

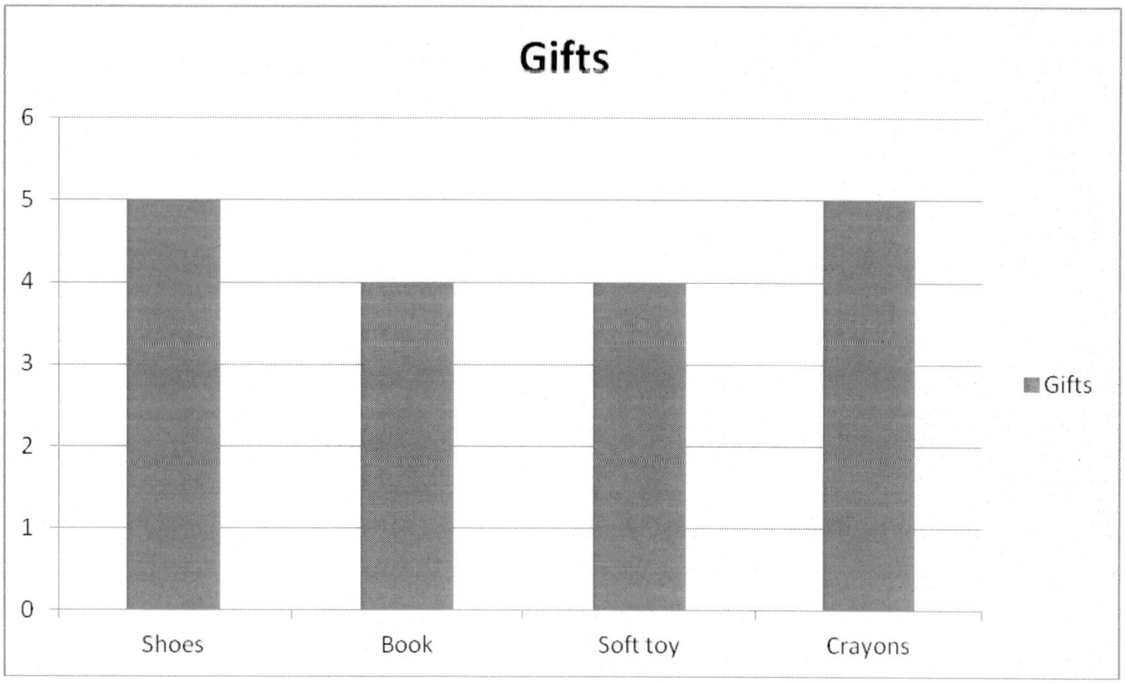

a) Which child's information is missing in the graph?

1 Mark

b) Draw the graph with all four children's information.

1 Mark

6. The diagram shows a square of side length 14cm.
Inside the square are 8 congruent trapeziums and a shaded square.

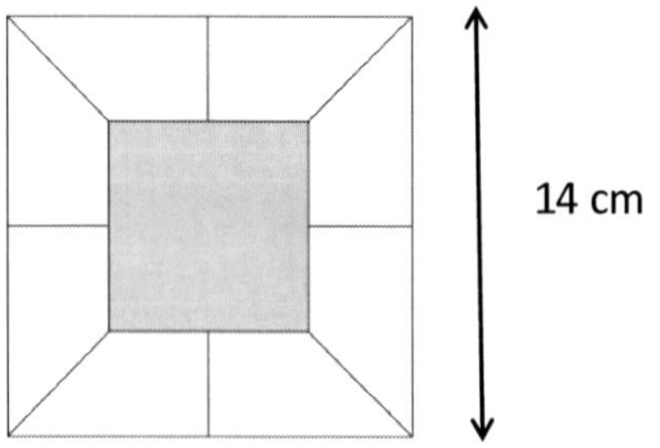

The area of a trapezium is 20cm^2. The height of a trapezium is 4 cm.

What is the area of the shaded square?

Show your working

2 Marks

7. Eighty children need 2 books and one pencil each.
There are 20 books in a box and 10 pencils in a bag.
How many boxes of books and bags of pencils are needed?

Show your working

2 Marks

8. The signs are missing from these number sentences.
Write the missing signs in the shaded boxes, **+, -, x and ÷.**

20	+	8	=	4	X	7
9		9	=	3		6
6		4	=	15		9

2 Marks

9. Round the following numbers.

758 to the nearest 100

956 to the nearest 10

$2\frac{3}{8}$ to the nearest whole number

97.3962 to the nearest hundredth place

2 Marks

10. Lara goes to shop to buy peanuts.
Each bag of peanut cost £2.79.
How many bags of peanuts can Lara Buy for £10?

Show your working

1 Mark

11. Sara has a basket that contains 12 oranges and 14 apples.

a) What fraction of the fruits in the basket are apples?

Sara gives one apple and one orange to Rita.

b) What fraction of the fruits in the basket are oranges now?

1 Mark

12. Below is a recipe for a cake.

250g butter, softened
1 cup caster sugar
2 teaspoons vanilla essence
3 eggs
2 1/2 cups self-raising flour
2/3 cup milk

Karen uses the recipe to make cake.
She uses 1 1/3 cups of milk.

How many eggs and vanilla essence will she use?

Show your working

1 Mark

13. The below graph shows a survey conducted among a group of children in order to find out children's favourite fruit.

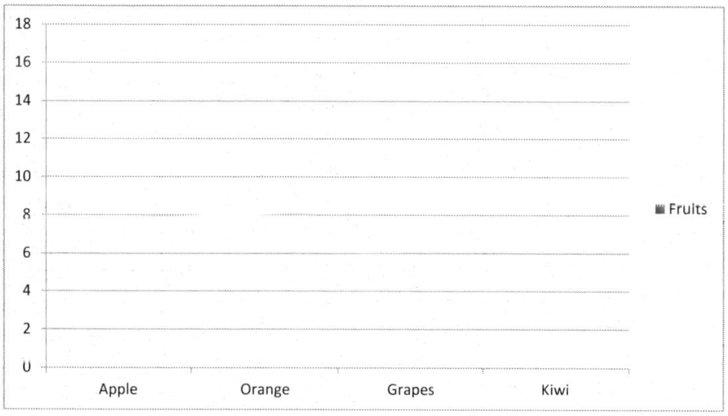

How many more children like oranges than grapes?

1 Mark

Rita says 'Exactly thrice as many children like oranges as children like kiwi'.

Is Rita correct? Yes/ No

Explain how you know?

1 Mark

14. Sara takes a taxi to go to her aunt's place.

It cost £1.50 for each kilometer.

She pays the taxi £12.75.

How many kilometers did she travel in the taxi?

Show your working

2 Marks

15. One double-digit number and two single-digit numbers multiply to make 576.

Write the missing numbers.

☐ x ☐ x ☐ = **576**

1 Mark

16. Rita and Sara have two 3D objects. Rita's object has 4 triangle faces. Sara's object has 4 triangle faces and one square. What are the names of the two objects?

Rita ☐

Sara ☐

1 Mark

17. Write the missing number.

(23 + ☐) x 28 = 1036

2 Marks

18. Amy decides to build a pond in her garden.
She orders 9000 bricks.
The store can deliver 750 bricks in a truck at a time.

a) How many times do they have to send the truck to Amy's place?

b) Amy returns 600 bricks. The cost of the balance bricks is £1260.
How much does a single brick cost?

2 Marks

19. Emily has eight 5p coins and fifteen 10p coins.
How much money does she have altogether?

1 Mark

20. Sara multiplies two numbers and divides the result by three.
Her answer is 32.
One of Sara's numbers is 8.

What is Sara's other number?

2 Marks

21.

Box of
cereal

Box of
Butter

Rita buys 6 boxes of cereal and 8 boxes of butter.

If she pays £10, how much balance will she get?

Show your working

2 Marks

22. Below shown is a grid with design.

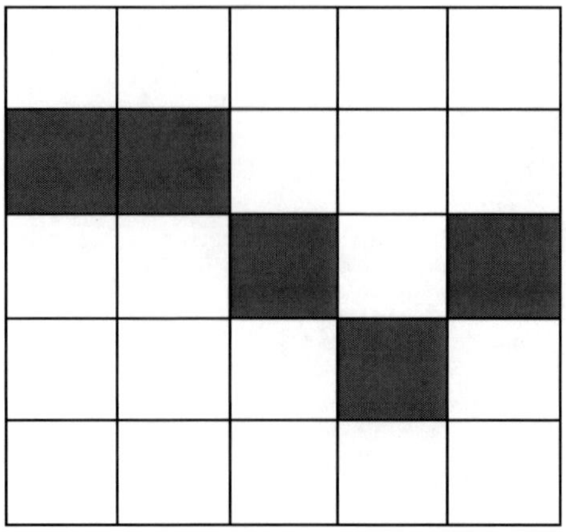

Sara rotates it to 90°. Shade in the missing parts of design.

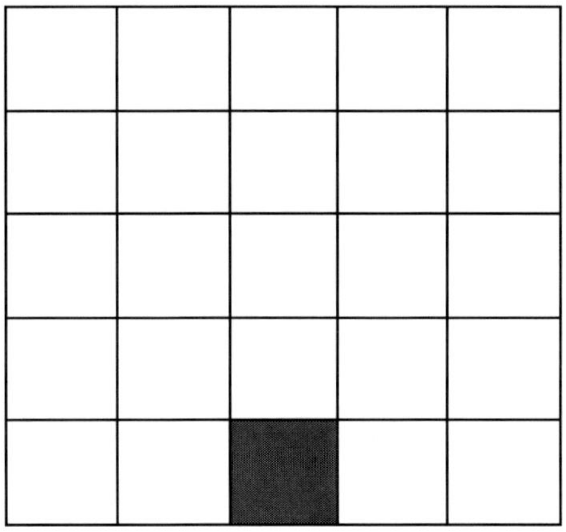

2 Marks

23. Rita has two equal size squares.

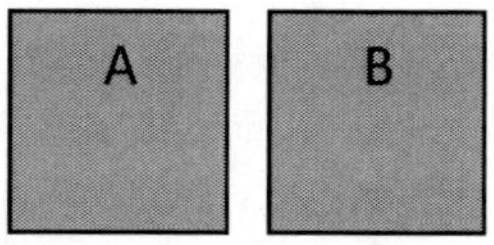

Each square's perimeter is 72cm.

She cuts square B in half and attaches it to square A to make a rectangle.

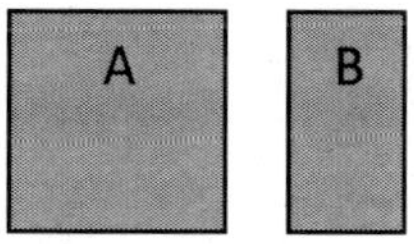

What is the perimeter of the rectangle?

Show your working

2 Marks

24. Cross (x) the shape that is exactly ¼ shaded

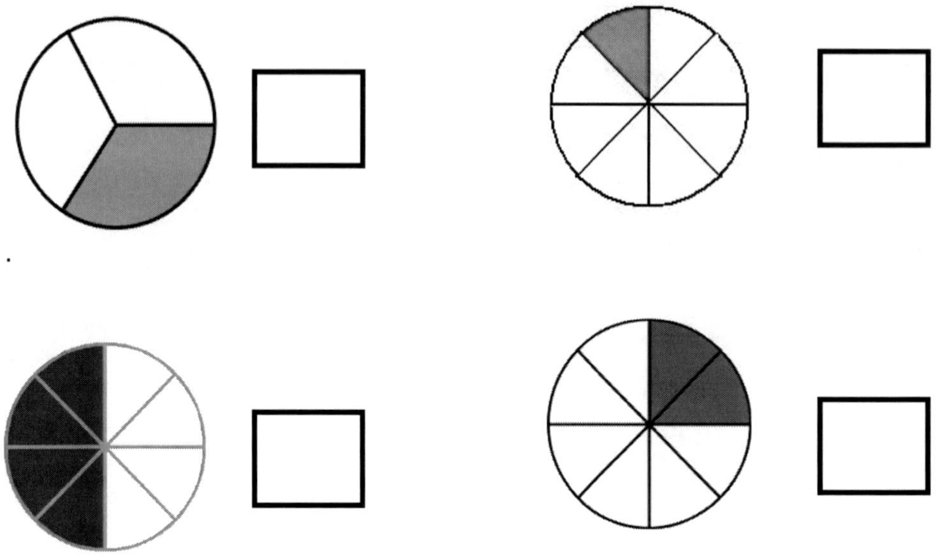

1 Mark

[END OF TEST]

Please do not write on this page.

Mathematics
Key stage 2

Paper 7

Calculator NOT allowed

For marker's use only

Page	Marks
Total	

Instructions

You **must not** use a calculator to answer any questions in this test.

Work as quickly and as carefully as you can.

You have **45 minutes** for this test.

If you cannot do one of the questions, **go on to the next one.**

You can come back to it later, if you have time.

If you finish before the end, **go back and check your work.**

Follow the instructions for each question carefully.

This shows where you need to put the answer.

If you need to do working out, you can use any space on a page.

Some questions have an answer box like this:

For these questions you may get a mark for showing your working.

1. Write these **prices** in order, starting with the **smallest**.

£45.50	£4.50	£4.55	£45

£	£	£	£

1 Mark

2. This table below shows six different types of tiger and where they are found in the Asian countries.

	India	China	Bangladesh
Bengal tigers	✓	✗	✓
Indochinese tigers	✓	✓	✓
Malayan tigers	✗	✗	✓
South china tigers	✗	✓	✗
Sumatran tigers	✗	✗	✓

Use this table to answer this question

Which type of tiger is **only** found in China?

Which types of tiger are found in **all** three parts of the countries?

2 Marks

3. Calculate 497 - 125

1 Mark

4. Here are five number cards.

$\frac{1}{2}$	4	$2\frac{1}{2}$	$4\frac{1}{2}$	7

Use three of the number cards to make this **calculation correct**.

(☐ + ☐) * ☐ = 49

1 Mark

5. Circle the number which is closer to 100.

46 56

Explain how you reached that conclusion.

1 Mark

6. Calculate 549 **x** 7

1 Mark

7. **Match** each shape to the correct name.
One has been done for you.

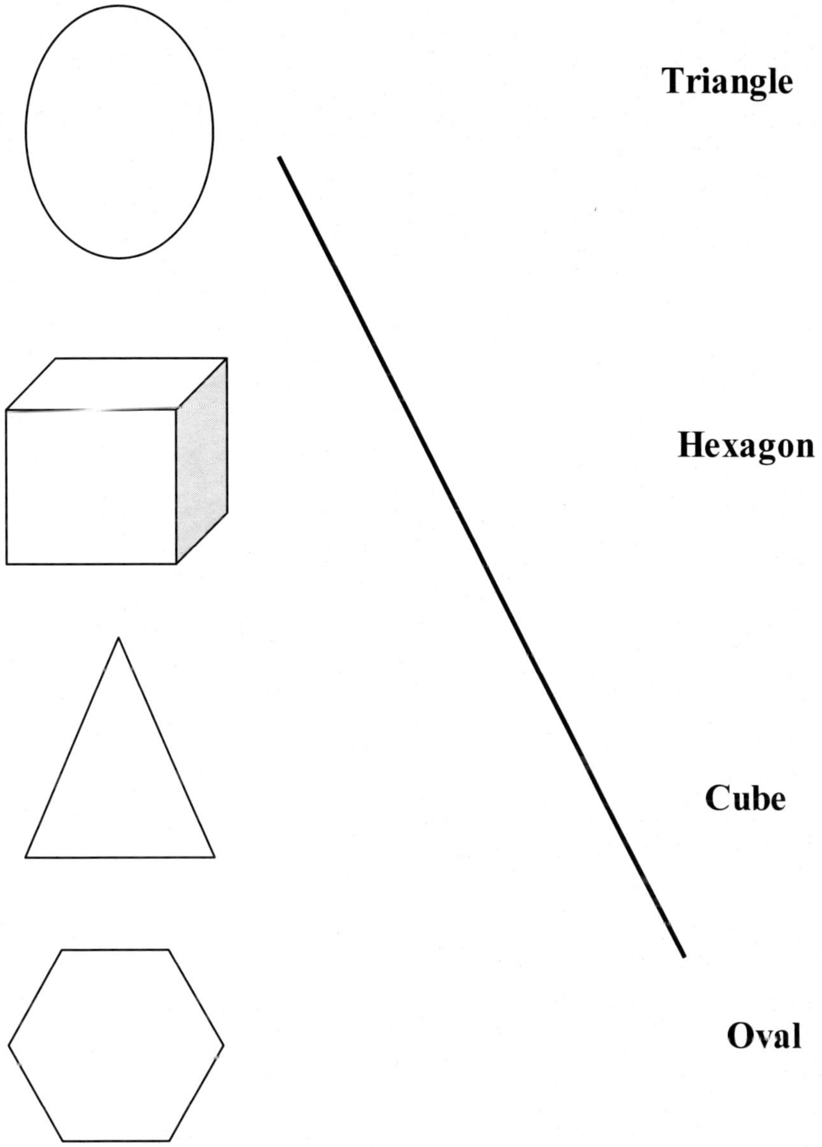

2 Marks

8. Liam makes a **sequence of numbers** starting with 500

He subtracts 150 each time.

Write the next two numbers in Liam's sequence.

500 350 200 ☐ ☐

2 Marks

9. **Circle** the fraction that is greater than ½ but less than ¾

9/11 **3/6** **2/3** **1/9** **16/20**

1 Mark

10. Calculate $\frac{2}{7}$ **of 350**

☐

1 Mark

11. Write the missing number to make this calculation correct.

1111 / ☐ = 11

1 Mark

12. Here is a number machine.

Here is another number machine.

Write the **four missing** answers **A, B, C and D.**

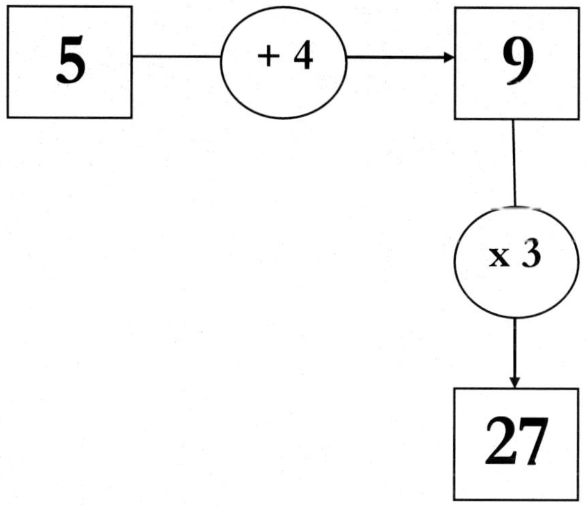

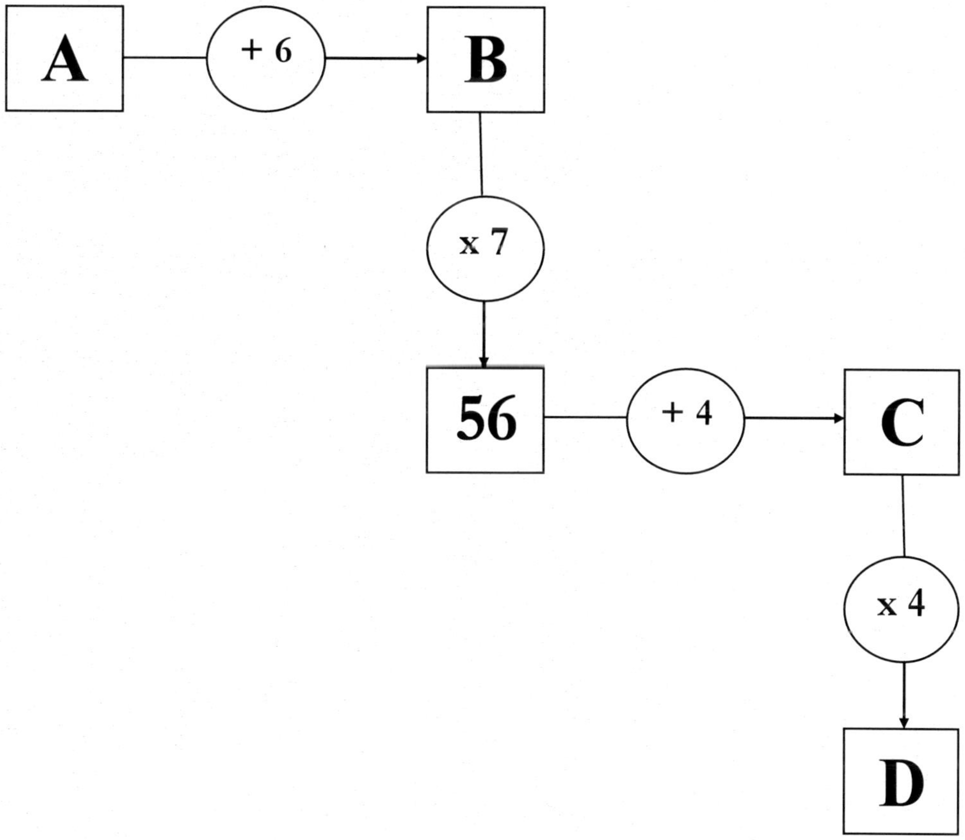

2 Marks

13.

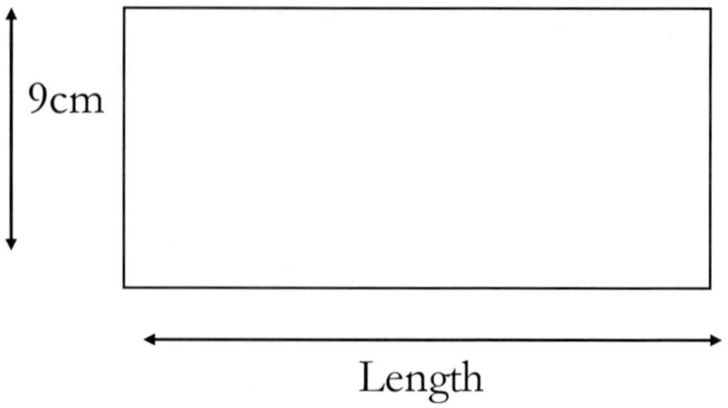

Not actual size

The perimeter of this rectangle is 60 centimeters. Find the length.

Show your working

2 Marks

14. Calculate $\frac{2}{5}$ **of 280**

2 Marks

15. Complete this table to show the numbers rounded to the **nearest 100**
One has been done for you.

	Rounded to the nearest 100
475	500
4955	
45565	
415289	

2 Marks

16. The tables show the cost of a new cricket kit.

Item	**Cost**
Bat	£ 49.75
Helmet	£ 37.50
Gloves (1 pair)	£ 12.85

Altogether, how much does the complete cricket kit cost?

1 Mark

17. A box of **16 chocolates** cost **£24.00.**

a). How much does **each** chocolate cost?

1 Mark

Lisa wants 6 chocolates and Mia wants 10 chocolates.

Lisa and Mia both pay some money towards the box of chocolates.

Lisa pays a third of the cost.

a) How much money does **Mia pay**?

Show your working

£

2 Marks

18. Alfie buys two dolls, each at the same price.

She pays with a £ 50 note and gets £ 17.50 change.

What is the cost of one doll?

Show your working

£

2 Marks

19. What is the **next two numbers** in the sequence below.

144 121 100

2 Marks

20. This graph shows the age group of tourists visiting a museum on a particular day.

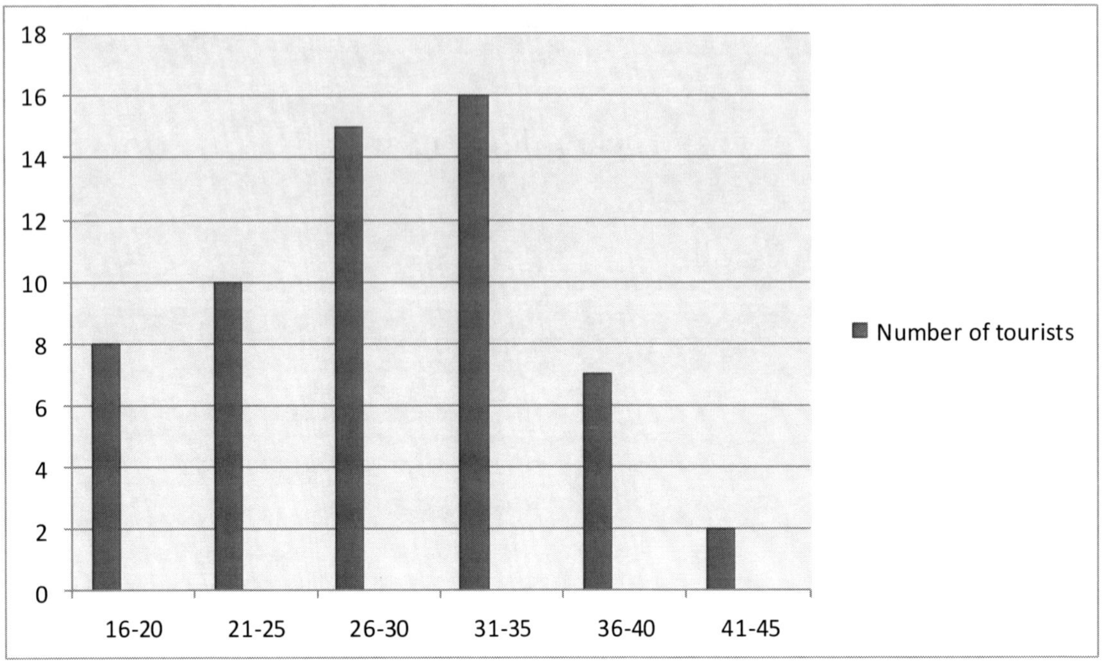

a. How many tourists are aged 30 or older?

1 Mark

b. A tourist aged 27 and a tourist aged 28 are not included in this graph. Add this information to the above graph.

1 Mark

21. Sarah, Lisa and Mia all go to the corner shop for breakfast.

a). Mia has **£5** to spend. She wants to buy a donut for £2.75 and a coffee for 1.80p. How much does she have left over?

Show your working

P

2 Marks

22. What is **40% of 150**

1 Mark

23. X says,

'All numbers that end in a 6 are multiples of 6'

Is he correct?

Circle Yes/No.

Yes **No**

Explain how you know

2 Marks

24. Circle two decimals that have a difference of 0.50

0.15 0.35 0.50 0.65 0.95

1 Mark

25. Calculate 2975/35

Show your working

2 Marks

[END OF TEST]

Please do not write on this page.

Mathematics
Key stage 2

Paper 8

Calculator NOT allowed

For marker's use only

Page	Marks
Total	

Instructions

You **must not** use a calculator to answer any questions in this test.

Work as quickly and as carefully as you can.

You have **45 minutes** for this test.

If you cannot do one of the questions, **go on to the next one.**

You can come back to it later, if you have time.

If you finish before the end, **go back and check your work.**

Follow the instructions for each question carefully.

This shows where you need to put the answer.

If you need to do working out, you can use any space on a page.

Some questions have an answer box like this:

For these questions you may get a mark for showing your working.

1. Join each of these calculations to the number that is **nearest** to the correct answer.

One has been done for you.

114 +339	800
437-119	400
40x19	500
12x12	100
789-357	300

2 Marks

2. Sarah did a survey on the number of students enrolled for different types of language courses.

She then drew this Venn diagram.

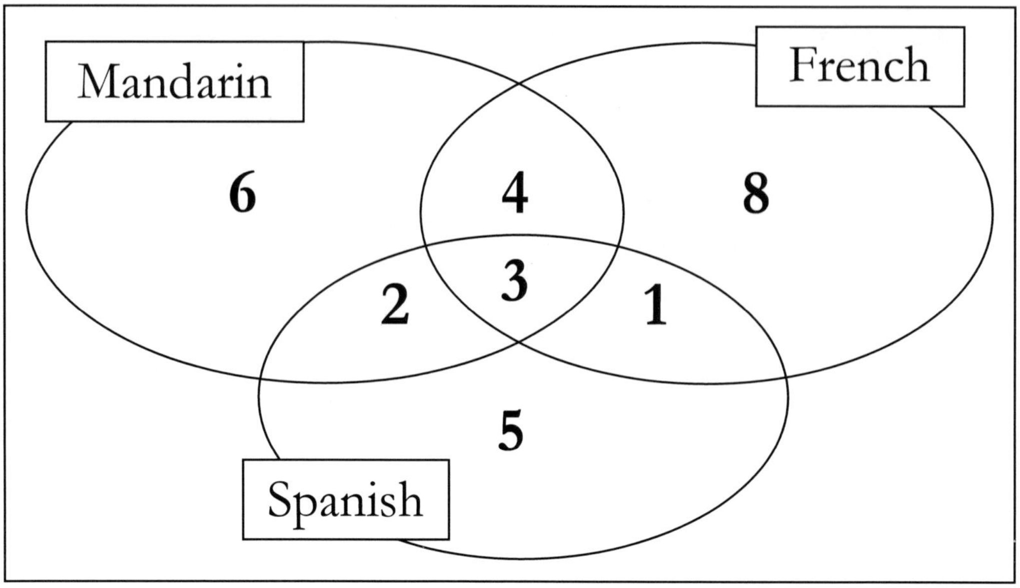

a). How many students followed French language class?

1 Mark

b). Which was the **most popular** language class?

Circle the correct answer below.

Mandarin French Spanish

1 Mark

3. What is 9/2 =

1 Mark

4. The square WXYZ is congruent to the square ABCD:

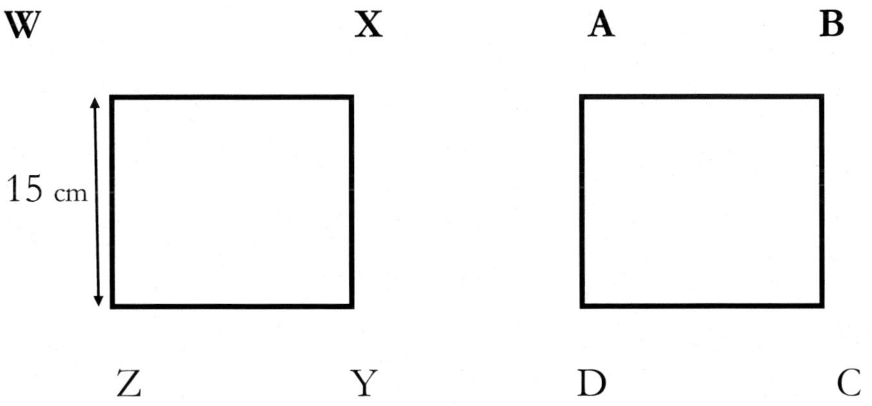

Find the following:

a. Length of AB =

b. Angle of C =

2 Marks

5.

Tickets for a cinema cost £5.75 each.

The management sold 17 tickets.

How much ticket money did the management collect?

1 Mark

6. How many edges does a hexagonal prism have?

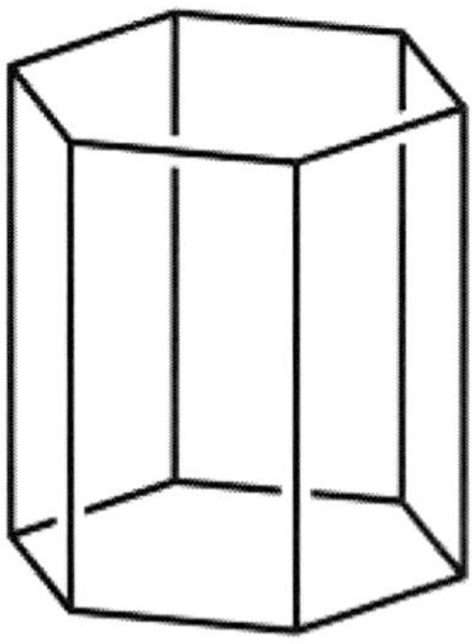

1 Mark

7. Calculate **55% of £390**

£

1 Mark

8. An apartment has 4 floors. The first floor has 10 rooms. Each of the other floors has 5 more rooms than the floor below it. How many rooms are in the apartment?

Show your working

2 Marks

9. Write the **missing number** in the box below.

X 30 = 510

1 Mark

10. Multiply the previous number by 3 then add 4.

Write in the **missing numbers** of the sequence.

□ 10 34 □ 322

2 Marks

11. A triangle has a height of 20 feet. The length of its base is 9 feet.

What is the area of the triangle?

□

1 Mark

12. Look at this diagram.

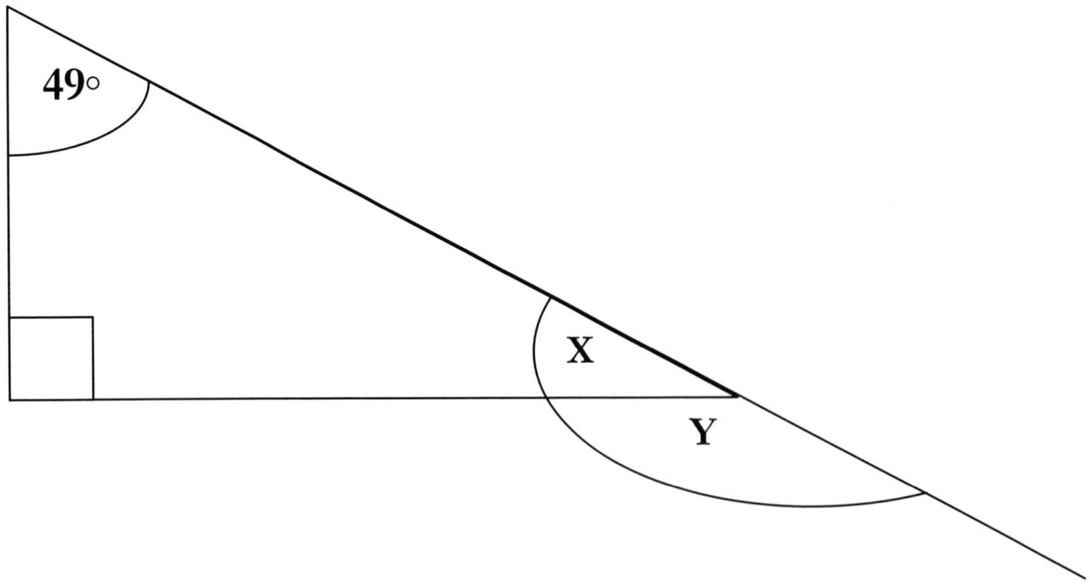

Calculate **angle x** and then calculate **angle y.**

Do not use a protractor.

a). **X** =

b). Y =

2 Marks

13.

The numbers in this sequence increase by 4 each time.

4 8 12 16 20

The numbers in this sequence increase by 6 each time.

6 12 18 24 30

Both sequences continue.

Write a number greater than 100 which will be in both sequences.

Show your working

2 Marks

14.

Lisa listed the prices of meals on a menu.

£ 5.25, £ 3.65, £ 7.50, £ 10.75, £ 9.20,

What is the mean price of the meals?

2 Marks

15.

X and Y each stands for a whole number.

X - Y = 360

X is **five times bigger than** Y.

Calculate the numbers for **X and Y.**

Show your working

X =

Y =

2 Marks

16.

An oil barrel needs 160,000 ml oil to fill its full capacity.

There is 5000 ml oil in a bottle.

How many bottle needed to fill the oil barrel?

2 Marks

17.

These are the opening times at London museum.

Monday to Thursday	11am to 4pm
Friday	Closed
Saturday	9am to 7pm
Sunday	11.30am to 8pm

How many hours is the London museum open on Saturday

1 Mark

18. Lisa covers the box shown with paper

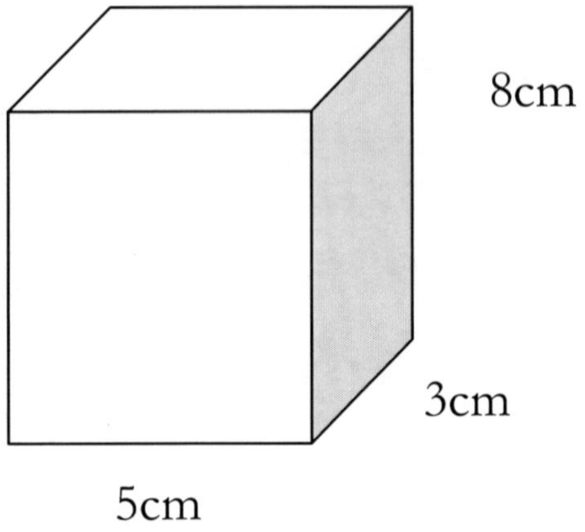

What is the surface area of the box?

Show your working

2 Marks

19.

A farmer has **500 cows** on his farm. At the market he **sells 60** of his cows and buys **18 sheep.**

How many animals does the farmer have on his farm **after he returns** from the market?

Show your working

2 Marks

20. Six children took part in the table tennis competition.

They each played five games.

This table shows how many games each child won, drew or lost.

	Won	Drew	Lost
Ann	2	2	1
Lisa	1	1	3
Mary	2	1	2
Mike	1	3	1
Shaun	0	4	1

How many children drew more games than they lost?

1 Mark

Each child scores two points for a win.

One point for a draw

No points for a loss

Who scored the most points?

1 Mark

21. What is the value of 4k +6 (j −2)

When k =3 j =5?

Show your working

2 Marks

22. Write the missing number.

[ ] - 54.58 = 157.23

1 Mark

23. A shop sells torches and batteries.

Mike bought three batteries and a torch.

The total cost was £16.90

The torch cost £10.00

How much did each battery cost

Show your working

2 Marks

[END OF TEST]

Please do not write on this page.

Mathematics
Key stage 2

Paper 9

Calculator NOT allowed

For marker's use only

Page	Marks
Total	

Instructions

You **must not** use a calculator to answer any questions in this test.

Work as quickly and as carefully as you can.

You have **45 minutes** for this test.

If you cannot do one of the questions, **go on to the next one.**

You can come back to it later, if you have time.

If you finish before the end, **go back and check your work.**

Follow the instructions for each question carefully.

This shows where you need to put the answer.

If you need to do working out, you can use any space on a page.

Some questions have an answer box like this:

For these questions you may get a mark for showing your working.

1. Round 4.7829 to the nearest hundredth.

1 Mark

2. Here are four digits cards.

2	4	6	3

Use any of the number cards to make this **calculation correct**.

(☐ ☐ + ☐ ☐) = 100

1 Mark

3. Find the area of the rectangle in square centimetres

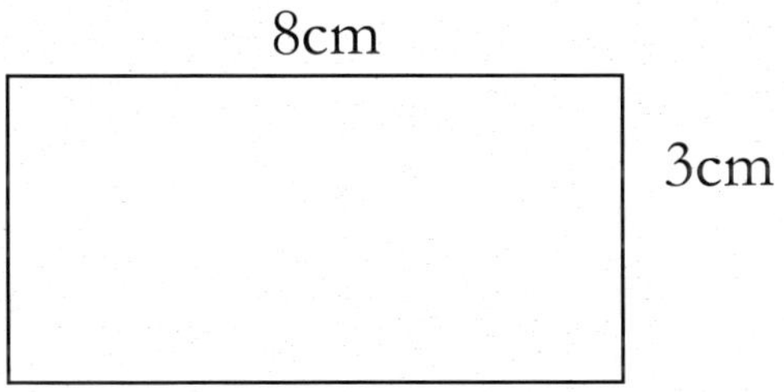

1 Mark

4. Write these times in order starting with the longest.

36 hours	2 ½ months
26 days	6 weeks

Longest

1 Mark

5. An oil tank has 50,000 liters, and 1,000 liters are taken out every day. What was the amount of oil in the tank 7 days ago?

Explain how you know?

2 Marks

6. Sarah watched a comic serial for 1 hour 10 minutes and a cartoon for 2 hour 7 minutes. Find the total duration which she watched TV.

1 Mark

7. A cylinder is given below

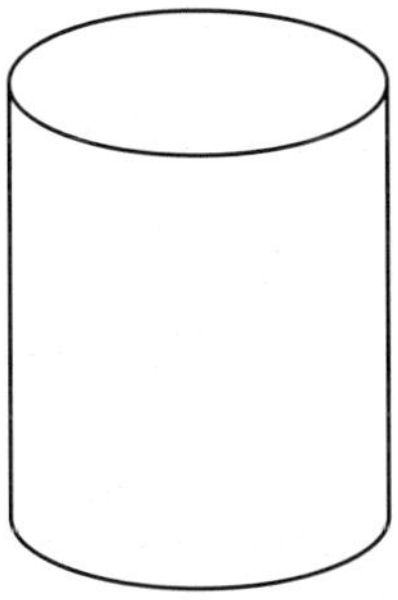

a. How many faces does it have

b. How many edges does it have

2 Marks

8. Lisa, Sarah & Sam made cookies to sell at their school fair.

They made 100 cookies altogether.

Lisa sold **13** cookies.

Sarah sold **35** cookies.

Sam sold **27** cookies.

How many cookies were left?

Show your workings here

2 Marks

9. Sarah makes a **sequence of numbers** starting with 64

Write the next two numbers in Sarah's sequence.

64 58 52 ☐ ☐

2 Marks

10. Circle the fraction of 0.85 in its lowest terms.

17/20 17/25 19/20 19/25

1 Mark

11. 100 bubble gums are shared equally among nine children.
How many bubble gums are left over?

1 Mark

12. Calculate 123 x 69

Show your working here

2 Marks

13. World population by continent in 2012.

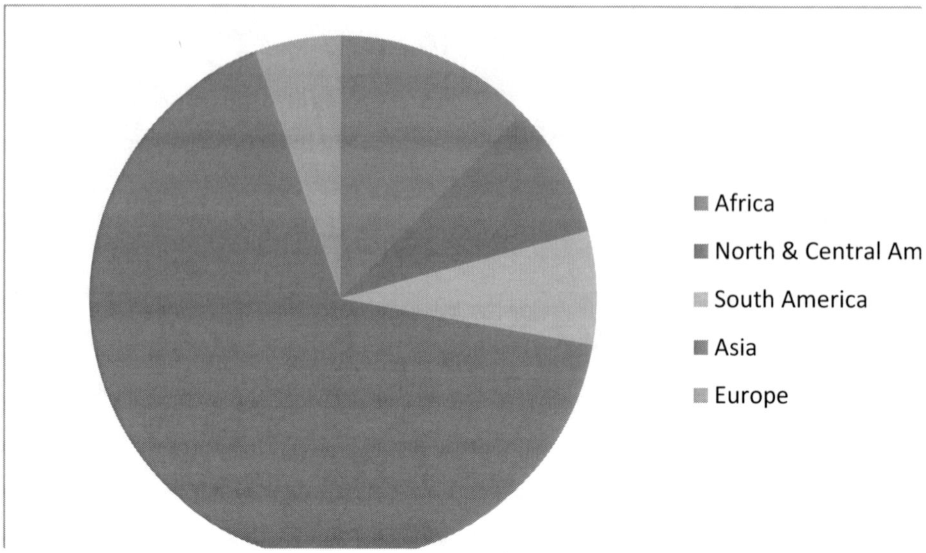

a. Circle the approximate percent of the world's population that live in Europe?

6%	22%	37%	45%

b. Circle the approximate percent of the world's population that live in Asia?

67%	33%	45%	53%

2 Marks

14. The area of this rectangle is 6 cm².

Calculate the base of this rectangle?

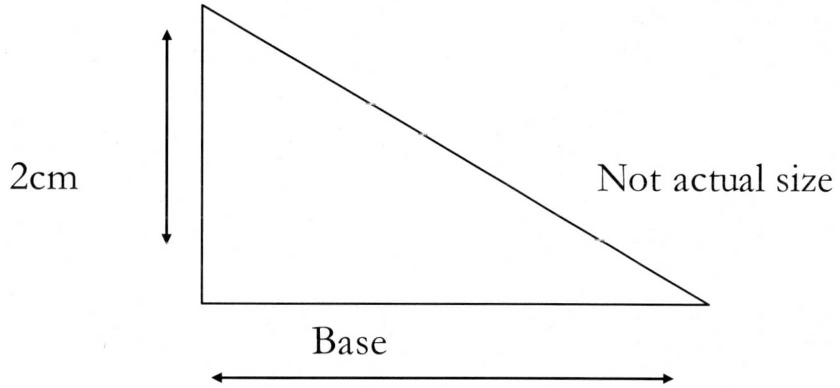

Show your working here

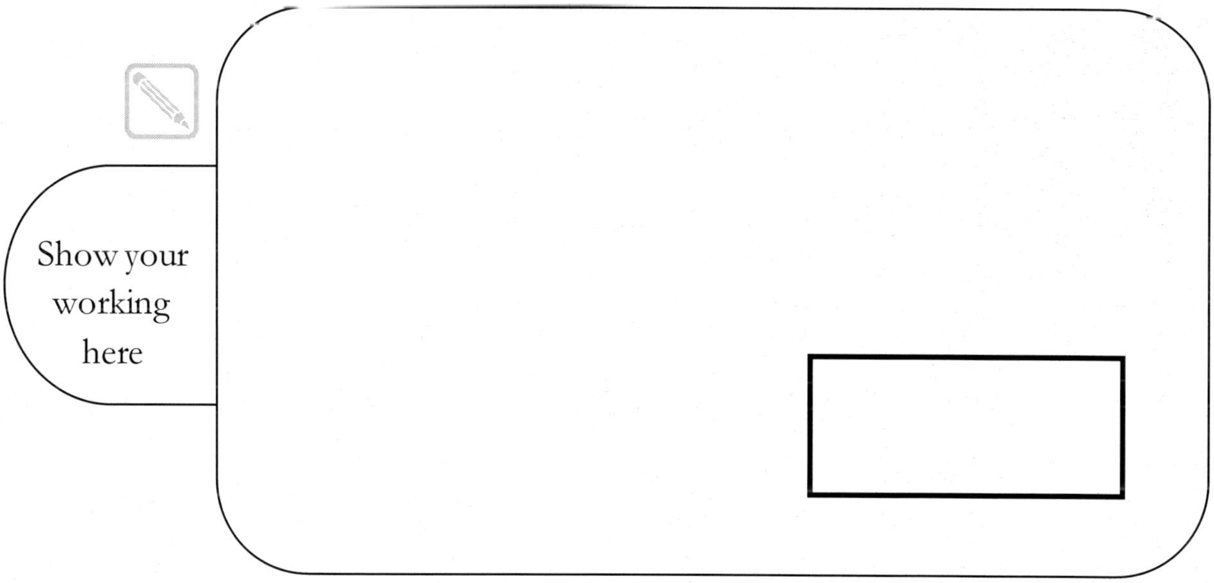

2 Marks

15. Add 3 ¾ and 2 ½

Show your working here

2 Marks

16. Part X & Y are joined by a straight line.
Draw a line to join point X on this sheet so that the two lines make a reflex angle.

X__________________________Y

2 Marks

17. The table shows the daily sales made by a mobile fast food company over one week.

Days	Sales (£)
Monday	510
Tuesday	630
Wednesday	370
Thursday	540
Friday	810
Saturday	1230
Sunday	1570

What was the value of the cumulative sales for Saturday?

2 Marks

18. The tables show the cost of toys.

Item	Cost
Toy car	£ 19.25
Toy boat	£ 57.50
Toy plane	£ 112.85

Altogether, how much does it cost to buy a toy plane and toy car?

1 Mark

19. A box of **24 matches** cost **£1.20.**

How much does **each** match cost?

1 Mark

20. Sarah has three boxes of gifts.

Box A & B together weigh the same as box C.

The three boxes weigh 1.2kg altogether.

Box A weighs 350g.

How much does box B weigh?

Show your working

2 Marks

21. Circle the correct answer.

The weight of pencil is about,

5g 50g 500g 5kg

1 Mark

22. This design has **one large square** and **two identical small squares**.
The design measures 54 centimeters by 42 centimeters.

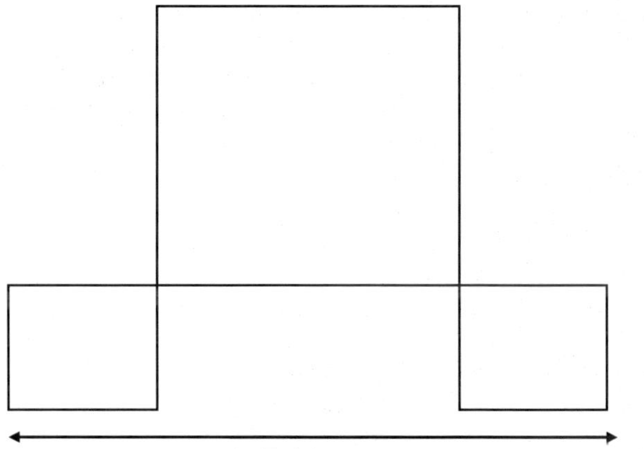

Calculate the length of a side of the large square.

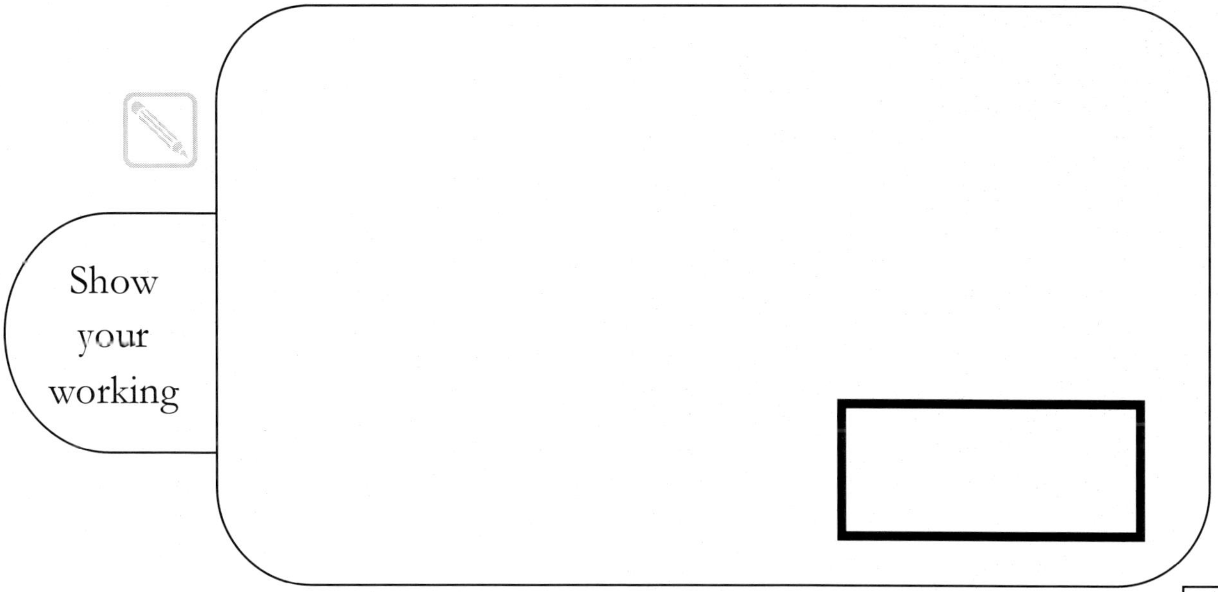

2 Marks

23. Sarah & Lisa each buy a book.
Sarah pays with a £10 note and gets £2.65 change.

Lisa book costs £12.50

How much more does Lisa's books cost than Lisa's book.

Show your working

2 Marks

24. What is **341 – 52 ½**

1 Mark

25. These are all times on the same morning.

A. 9:56am
B. quarter to nine
C. six minutes to nine
D. half past nine

Write the letters for the times in order starting with the earliest.

1 Mark

26. If you know 20% of a number explain how you could work out the original number.

Show your working

2 Marks

[END OF TEST]

Please do not write on this page

Mathematics
Key stage 2

Paper 10

Calculator NOT allowed

For marker's use only

Page	Marks
Total	

Instructions

You **must not** use a calculator to answer any questions in this test.

Work as quickly and as carefully as you can.

You have **45 minutes** for this test.

If you cannot do one of the questions, **go on to the next one.**

You can come back to it later, if you have time.

If you finish before the end, **go back and check your work.**

Follow the instructions for each question carefully.

This shows where you need to put the answer.

If you need to do working out, you can use any space on a page.

Some questions have an answer box like this:

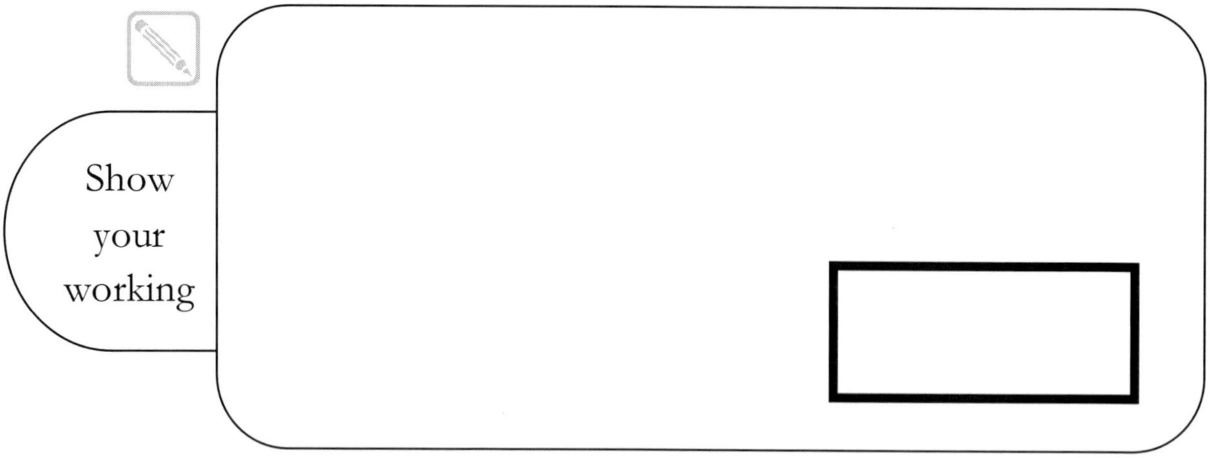

For these questions you may get a mark for showing your working.

1. Tick ✓ each division that has an answer greater than 100

One has been done for you.

1500 ÷ 5 = ✓

5500 ÷ 11 = ☐

750 ÷ 8 = ☐

225 ÷ 3 = ☐

18500 ÷ 175 = ☐

2 Marks

2. Calculate **79 x 96**

1 Mark

3. Arrange 3/4, 3/9, 3/7 and 3/10 in ascending and descending order.

a. Ascending order

b. Descending order

2 Marks

4. A bag contains 75kg of sugar. Find the quantity of sugar in 54 such bags.

1 Mark

5. The histogram shows the heights of 18 mountains in the world, grouped in to 500 meter height.

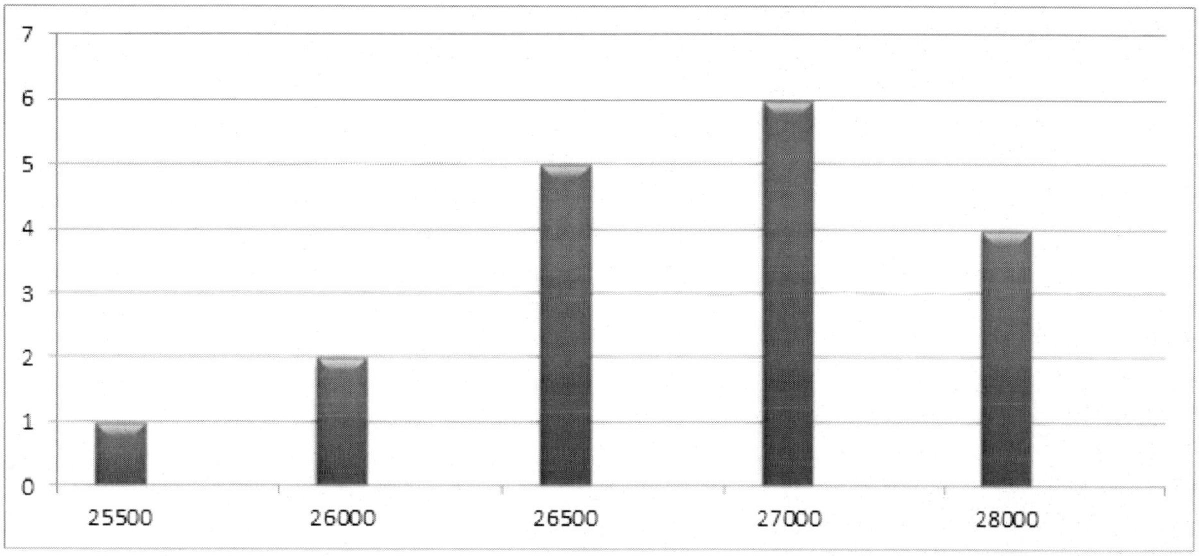

a. How many mountains were greater than or equal to 27000 meter height?

1 Mark

b. How many mountains were greater than or equal to 26000 meter height but less than 27000 meter height?

1 Mark

6. Circle the correct answer; find the numbers which are divisible by 13 without a remainder.

26 45 65 70 170

2 Marks

7. A shop sells coffee maker items.

Coffee bag	Sugar bag	coffee machine
		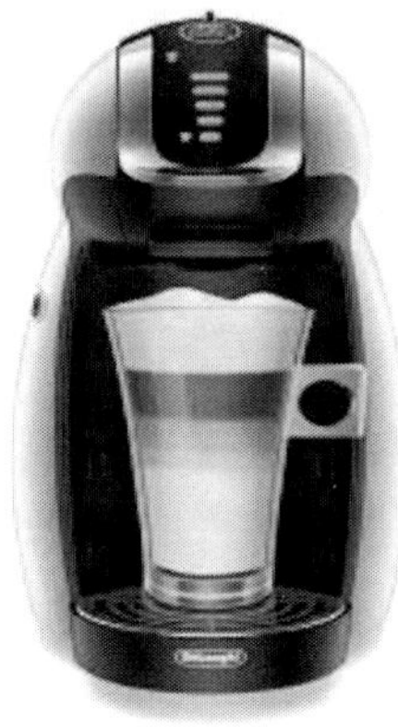
£13.79 per bag	£5.35 per bag	£68.95 each

Sarah has £150 to spend on coffee.

a. How many bags of coffee she can get for £150

1 Mark

Sam has £50.

He wants to buy a coffee machine, 3 bags of coffee and 2 bags of sugar.

b.How much more money does he need?

Show your working

2 Marks

c. Express 15/25 as a decimal.

1 Mark

d. Sarah walks for 67,500 m. How many kilometers is that?

1 Mark

e. For the formula $a^2 = b^2 + c^2$, what is the value of a when b = 7 and c = 24?

Show your working

2 Marks

f. A survey was done to find out students favourite book genres. This chart shows the results.

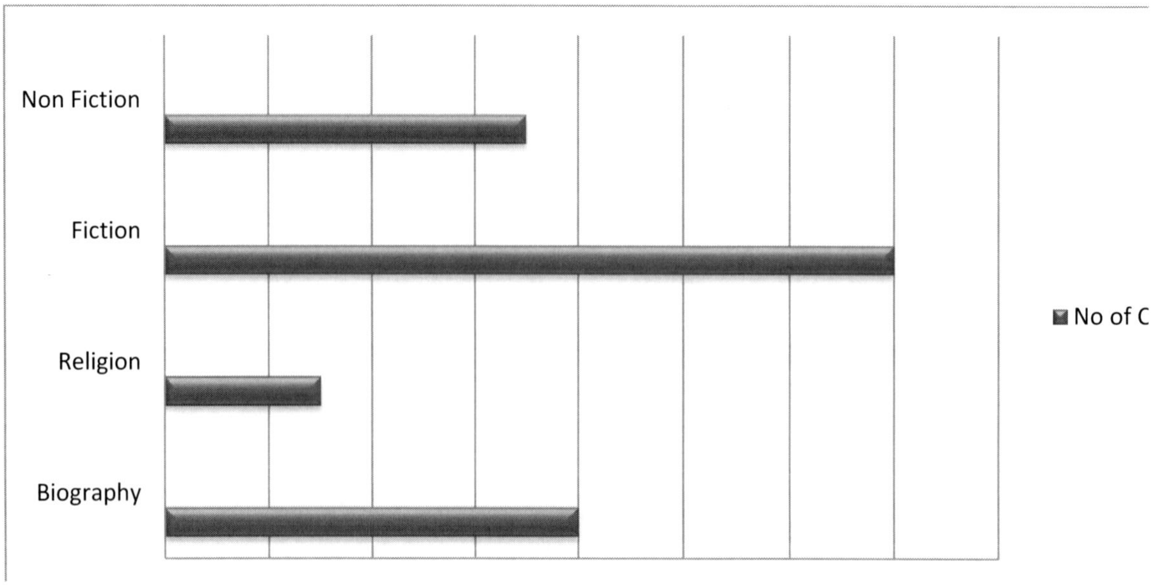

How many more children choose fiction rather than biography?

1 Mark

Sam says, "Exactly twice as many children choose fiction as non-fiction".

Is Sam correct?

Circle Yes No

Explain how you know?

2 Marks

g. Here is a digital clock.

00.17

What is 00:17 on the AM/PM clock?

1 Mark

h. **Match** each shape to the correct information.
One has been done for you.

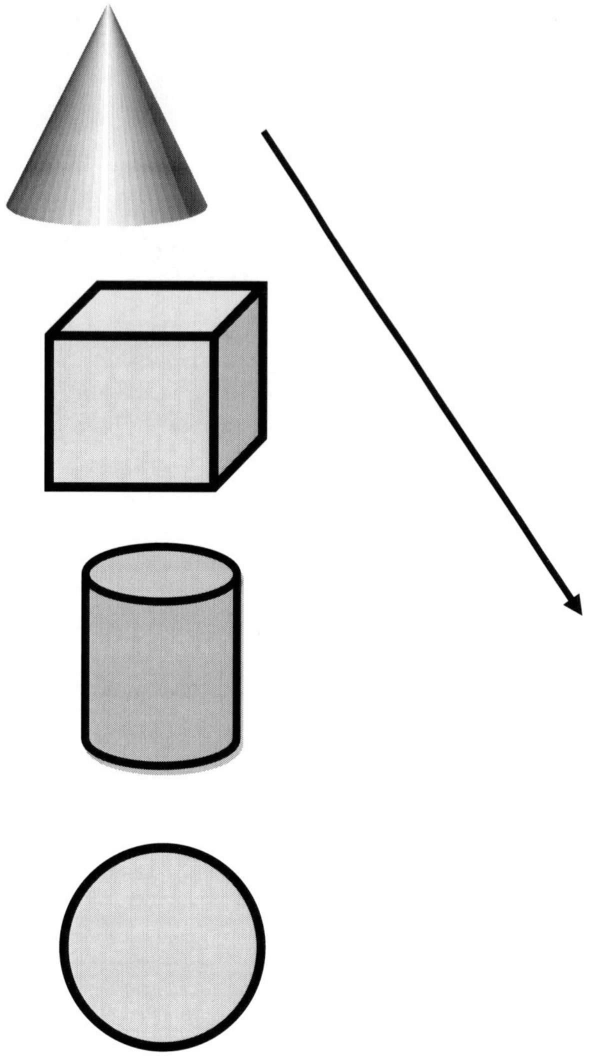

I have 12 equal edges.

I have no edges.

I have 1 round edge

I have 2 round edges

2 Marks

Find the value of x. write your answer as a decimal number.

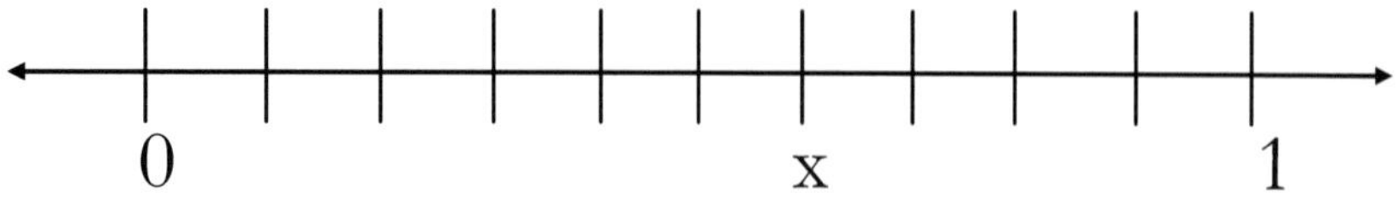

1 Mark

i. Write the two missing digits in this multiplication.

	7	X		7	= 459

2 Marks

j. What is the value of 5 + 16 ÷ 4 × 2?

2 Marks

k. By using the three letters on the shape that define the angle, angle x is written as:

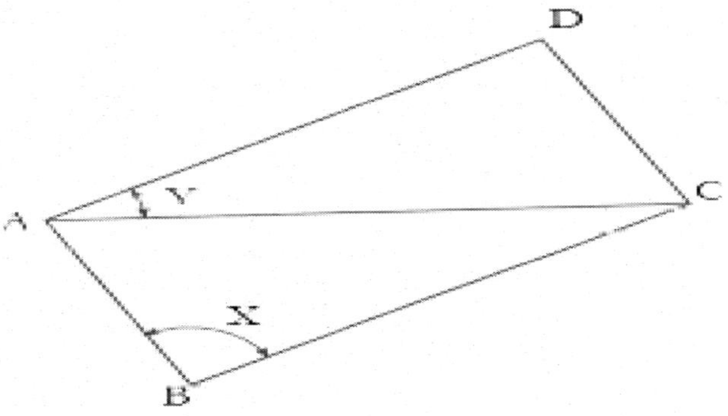

2 Marks

8. **Circle** the number that is closest to 2800.

2700 2740 2760 2860 2900

1 Mark

9. Estimate the square root of 50 to one decimal. Circle the correct answer.

a. 6.9

b. 7.9

c. 7.1

d. 8.4

Explain how you know?

1 Mark

10. A circular swimming pool has a radius of 14 m. What is it circumference?

Use 22/7 as an approximate for π

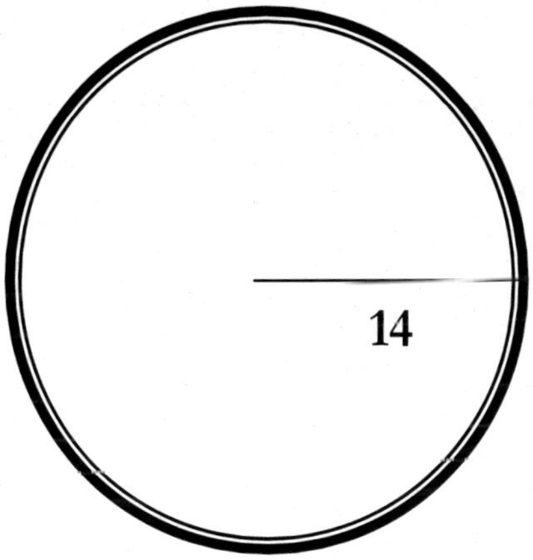

Show your working

2 Marks

11. Sam goes cycling every Saturday.

He went cycling on Saturday February 1st.

a. Altogether how many times did Sam go cycling in February

1 Mark

b. Sam travels the first 3 hours of his journey at 60 mph speed and the remaining 5 hours at 24 mph speed. What is the average speed of Jim's travel in mph?

2 Marks

12. There are 10 balloons in a bag: 5 are red, 2 are black and 3 are green.

The contents of the bag are shaken before Sara randomly chooses one balloon from the bag.

What is the probability that she **doesn't** pick a black balloon?

2 Marks

13. The number **9** goes into two of the square of this multiplication grid.

Tick (✓) the two squares where the 9 goes.

X	1	2	3	4	5
1					
2					
3					
4					
5					

2 Marks

[END OF TEST]

Please do not write on this page

Mathematics
Key Stage 2

Paper 11

Calculator NOT allowed

For marker's use only

Page	Marks
Total	

Instructions

You **must not** use a calculator to answer any questions in this test.

Work as quickly and as carefully as you can.

You have **45 minutes** for this test.

If you cannot do one of the questions, **go on to the next one.**

You can come back to it later, if you have time.

If you finish before the end, **go back and check your work.**

Follow the instructions for each question carefully.

This shows where you need to put the answer.

If you need to do working out, you can use any space on a page.

Some questions have an answer box like this:

For these questions you may get a mark for showing your working.

1. Write these prices in order, starting with the smallest.

32.50	3.50	5.30	30.50

£	£	£	£

1 mark

2. Calculate **460 – 192**

1 mark

3. Calculate **526 divided by 2**

1 mark

4. Laura, Mia and Annabelle made cupcakes to sell at their school bake sale.
They made **90 cupcakes altogether**.

Laura **sold 9** cupcakes.

Mia **sold 20** cupcakes.

Annabelle **sold 18** cupcakes.

How many cupcakes were left over?

Show your working

2 marks

5. Mia sold hers for **20p each**.
How much did she collect?

£

2 marks

6. **Match** each shape to the correct name.
   One has been done for you.

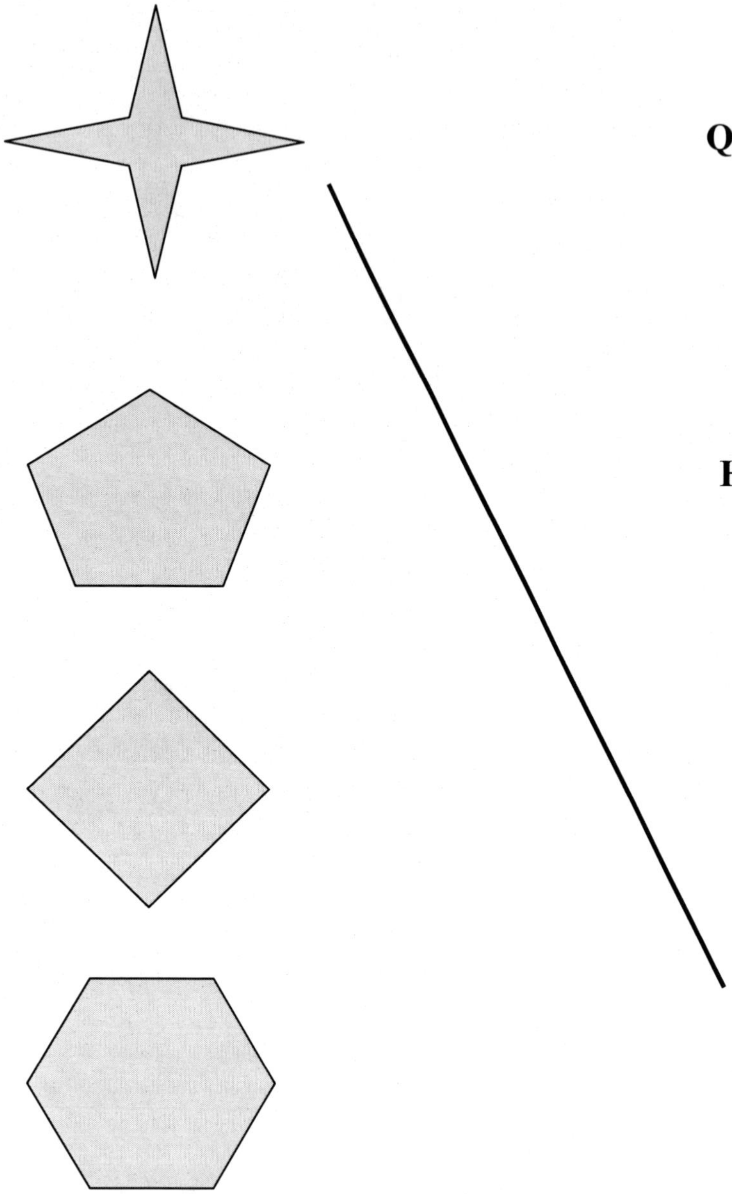

Quadrilateral

Hexagon

Pentagon

Octagon

2 marks

7. **Circle** the number which is **closer to 500.**

496 **506**

1 mark

8. Here is a number machine.

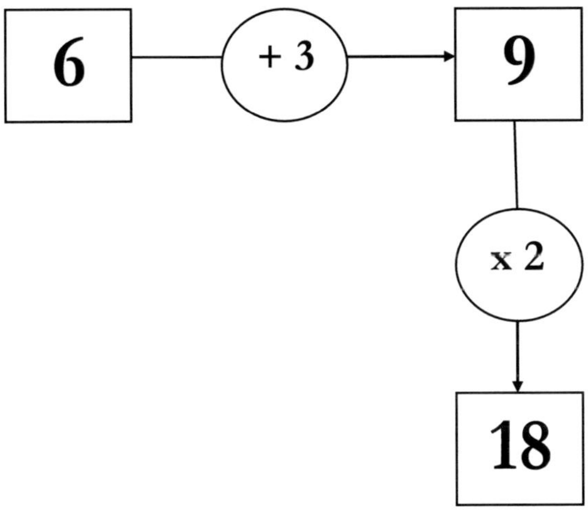

Here is another number machine.

Write the **four missing** answers **A, B, C and D.**

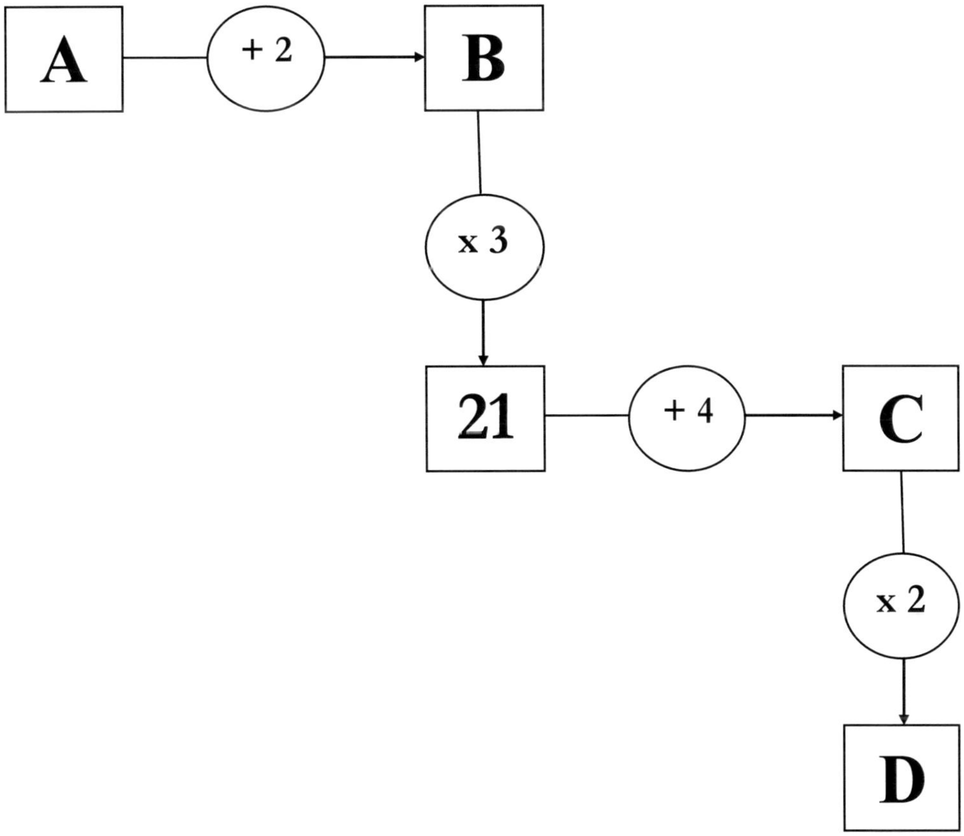

2 marks

9. This is the time (on the right) that Lisa left her house to walk to school.
The walk takes **25 minutes**.

What time did Lisa **arrive** at school?

1 mark

10. **Circle two** numbers that add up to make **0.92**.

0.53 **0.15** **0.08** **0.39** **0.21**

2 marks

11. Calculate $\frac{2}{5}$ **of 280**

2 marks

12. Calculate **732** ÷ **6**

1 mark

13. At the supermarket there are ice lollies for sale in **boxes of six**.

Sarah needs **24 ice lollies** for her birthday party tonight.

a). How many **boxes** of ice lollies does Sarah need?

boxes

1 mark

On the way home $\frac{1}{6}$ of the ice lollies melt.

b). How many ice lollies does Sarah **end up with** for the party?

1 mark

14. John has **£30.**

He buys **two trousers and a t-shirt.**

How much does he have **left**?

Show your working

£

2 marks

15. Match the wall clocks with the correct digital times.
An example has been shown for you.

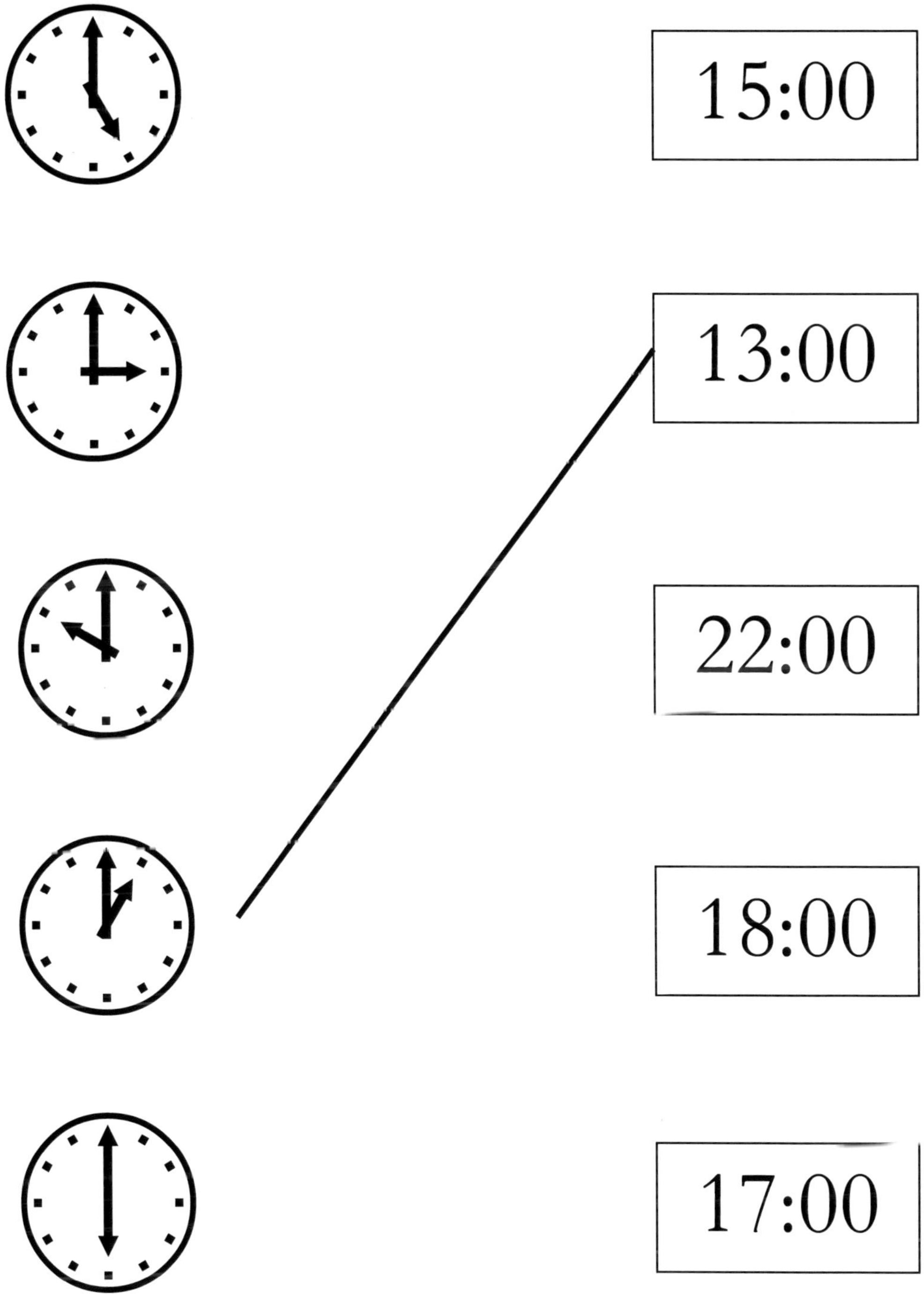

2 marks

16. A box of **9 donuts** cost **£7.20.**

a). How much does **each** donut cost?

p

1 mark

Lisa wants 3 donuts and Mia wants 6 donuts.

Lisa and Mia both pay some money towards the box of donuts.

Lisa pays a third of the cost.

b). How much money does **Mia pay**?

Show your working

£

2 marks

17. Circle the numbers which **divides by 4 with no remainder.**

32	52	22
23	16	36
63	31	53
56	28	66

2 marks

18. Mia chooses **two cards** from the four shown below.

12	9	24	6

She multiplies the numbers on both her cards together.

She then rounds the result to the nearest 10.

Her answer is 110.

Which **two cards** did Mia choose?

[ ] and [ ]

1 mark

19. Write these in order starting with the longest.

2 weeks	6 months	72 hours	2 days

longest

2 marks

20. Calculate **230 x 3**

1 mark

21. What is the **next two numbers** in the sequence below.

160 148 136

1 mark

22. Write the **missing number** to make the calculation correct.

12 x [ ] = 1212

1 mark

23. Sarah, Lisa and Mia all go to the corner shop for lunch.

a). Mia has **£3.50** to spend. She wants to buy a sandwich for £1.75 and an apple for 80p. How much does she have left over?

Show your working

[ ] **p**

2 marks

b). Sarah and Lisa share their lunch and they both have **£10.50** together. They buy 2 sandwiches for £1.75 each, a box of donuts for £4.50 and 2 lollipops for 20p each. How much money did they spend?

Show your working

£

2 marks

24. What is **40% of 120**

1 mark

[END OF TEST]

Please do not write on this page.

Mathematics
Key Stage 2

Paper 12

Calculator NOT allowed

For marker's use only

Page	Marks
Total	

Instructions

You **must not** use a calculator to answer any questions in this test.

Work as quickly and as carefully as you can.

You have **45 minutes** for this test.

If you cannot do one of the questions, **go on to the next one.**

You can come back to it later, if you have time.

If you finish before the end, **go back and check your work.**

Follow the instructions for each question carefully.

This shows where you need to put the answer.

If you need to do working out, you can use any space on a page.

Some questions have an answer box like this:

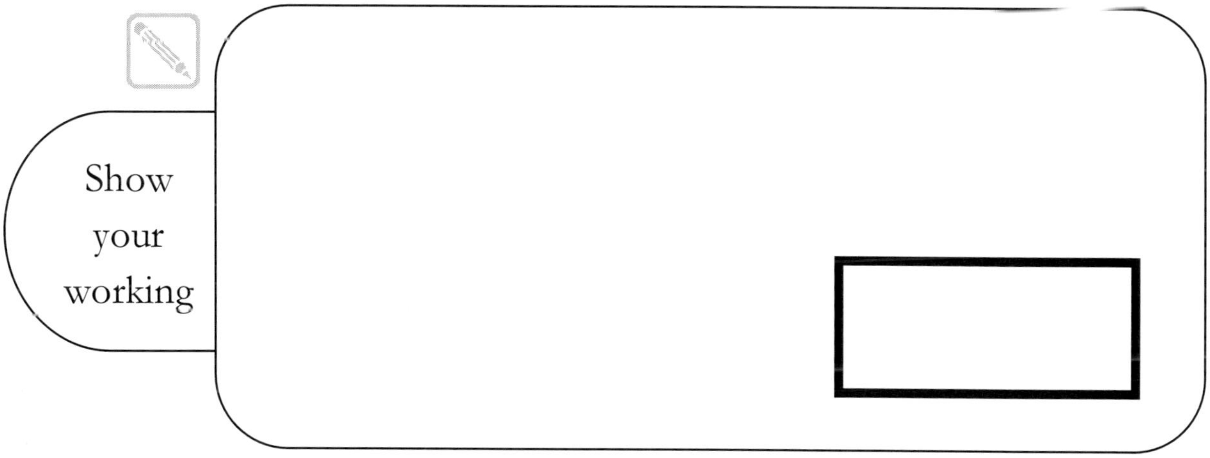

For these questions you may get a mark for showing your working.

1. Write the **missing numbers** in the calculations below.

a) 4.2 **x** ☐ = 4200

b) 5060 ÷ 100 = ☐

2 marks

2. Sarah did a survey on the number of pets her year group had.

She then drew this Venn diagram.

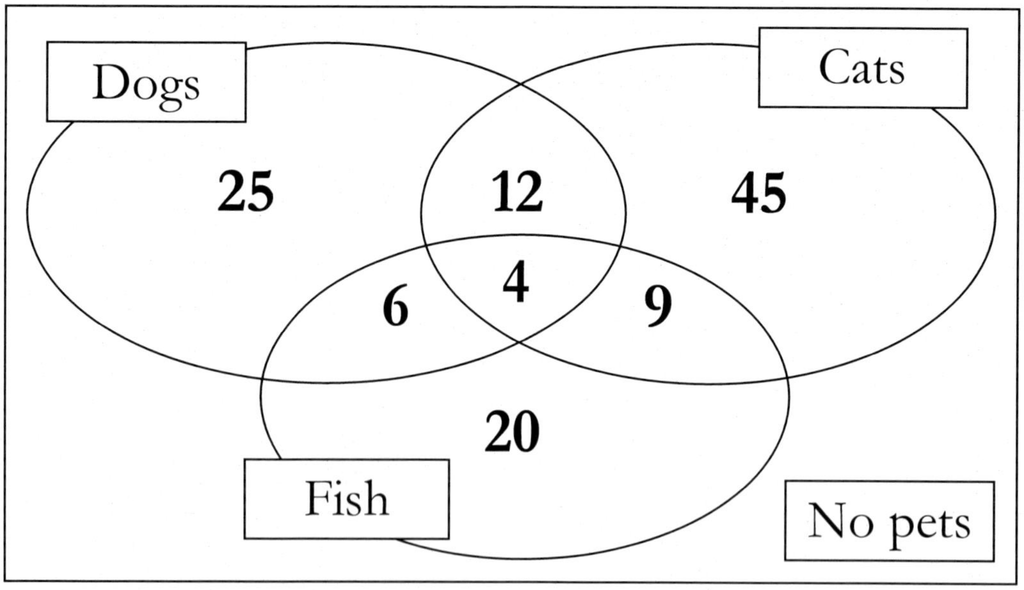

a). How many children had **a cat** for a pet? ☐

1 mark

b). Which was the **most popular** pet?

Circle the correct answer below.

DOG CAT FISH

1 mark

3. What is $\frac{3}{5}$ as a decimal

1 mark

4. Mia measured her friends right arm and tabulated the answers.

Complete the table.

	cm	m
Liam	149	
Visha	136	
Jack		1.50
Leana	143	1.43

2 marks

5. **Complete** the diagram below to make a shape that is **symmetrical** about the mirror line.

Use a ruler. You may also use a mirror or tracing paper.

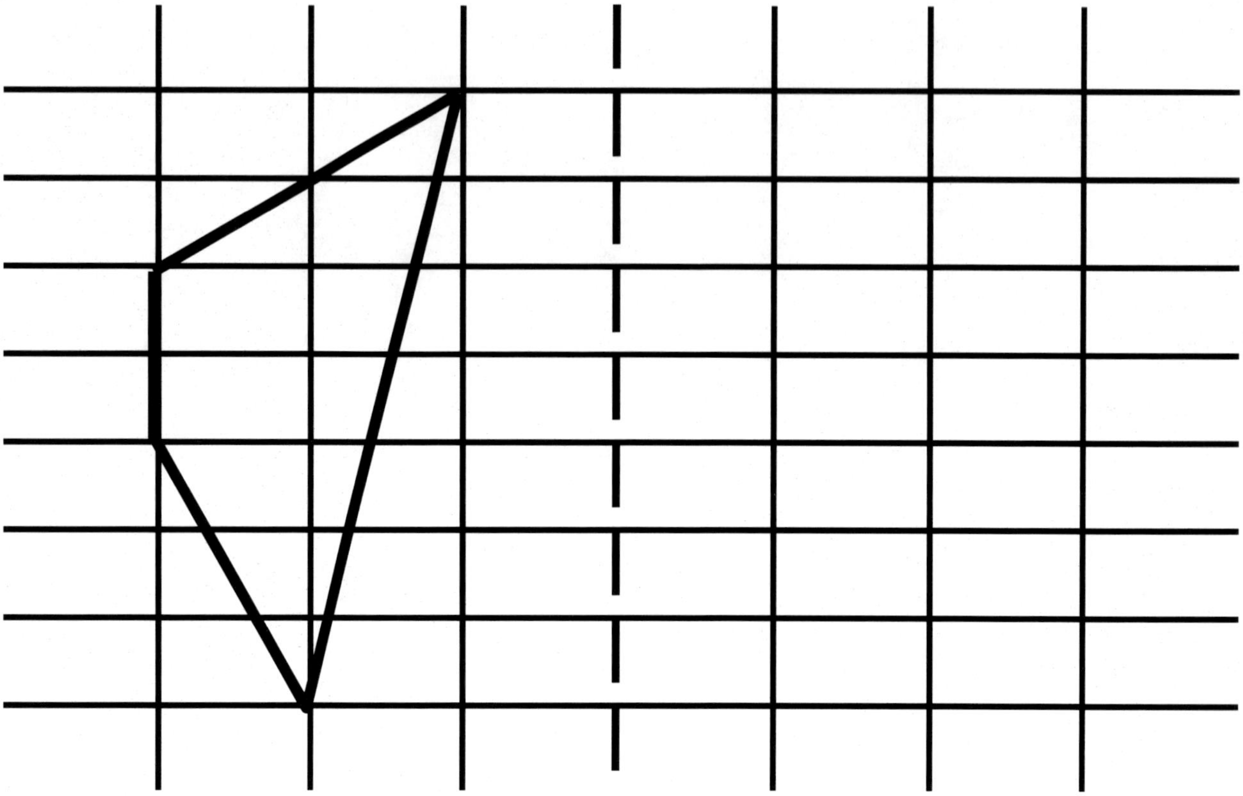

Mirror line

2 marks

6. Calculate **45% of £670**

£

2 marks

7. A flight to Canada lasts **9 hours.**

Lisa and her friends travel from London to Canada on a school trip.

The **time difference is 5 hours**. For example, if its 3 am in Canada, it will be 8 am in London.

Lisa's flight leaves London Heathrow airport at 11 am.
What time do they land in Canada?

Show your working

2 marks

8. Write the **missing number** in the box below.

$\square - 30 = 525$

1 mark

9. Multiply the previous number by 2 then add 5.

Write in the **missing numbers** of the sequence.

☐ 9 23 ☐ 107

2 marks

10. Here is a rectangle drawn on a graph.

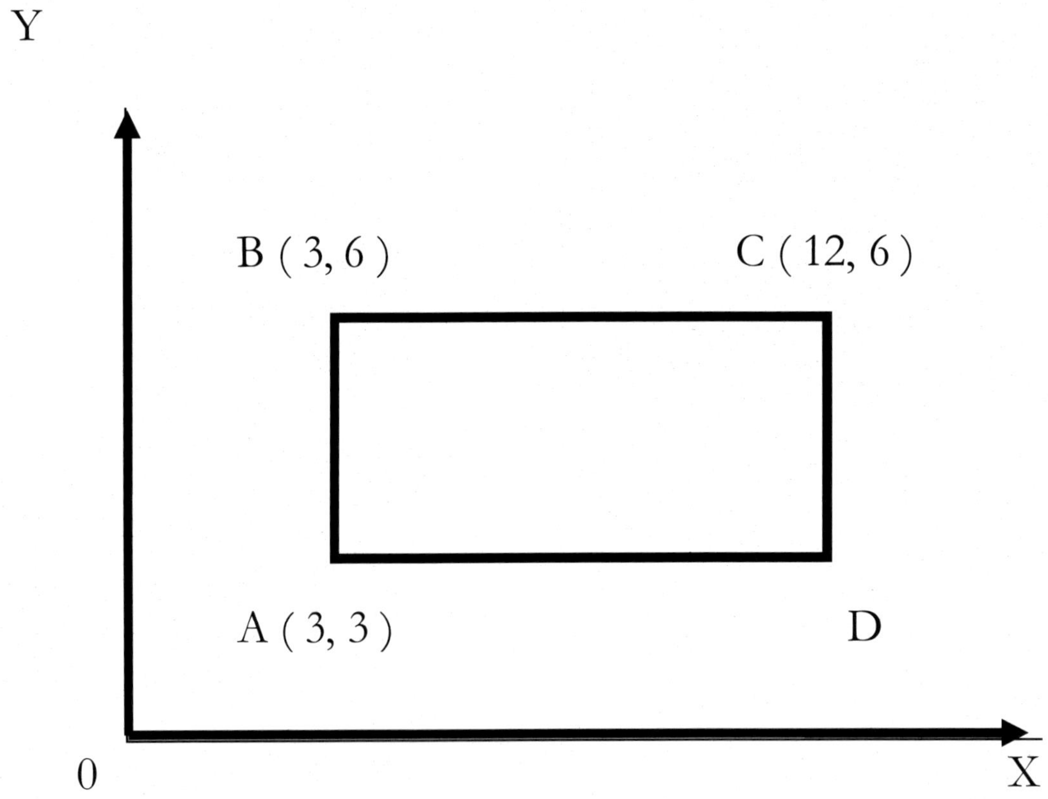

What are the **coordinates of point D**?

(☐ , ☐)

2 marks

11. Look at this diagram.

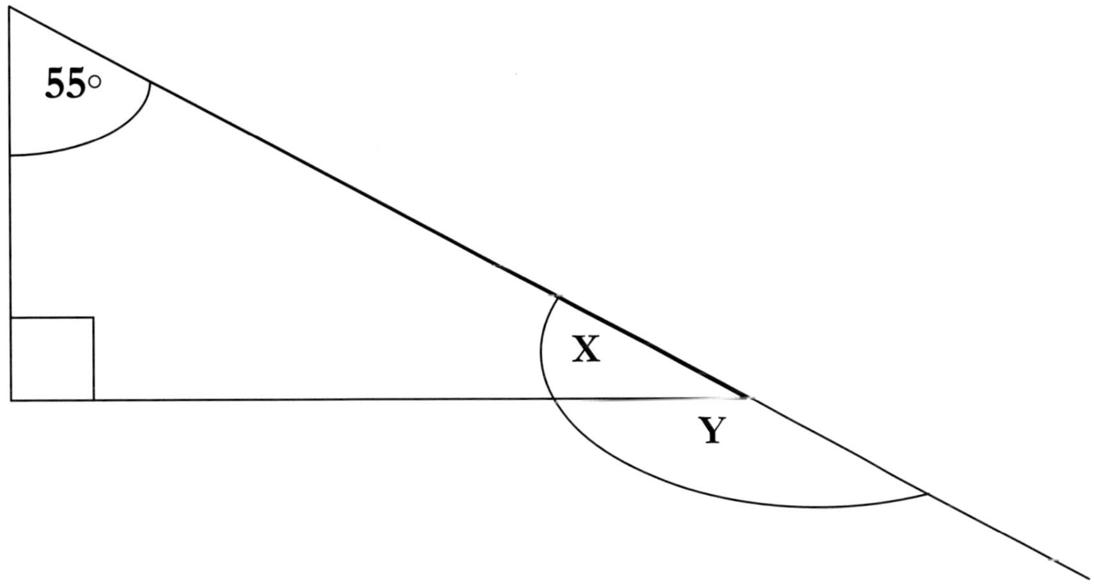

Calculate **angle x** and then calculate **angle y.**

Do not use a protractor.

a). **X** =

b). Y =

2 marks

12. Here are 3 shapes on a square grid.

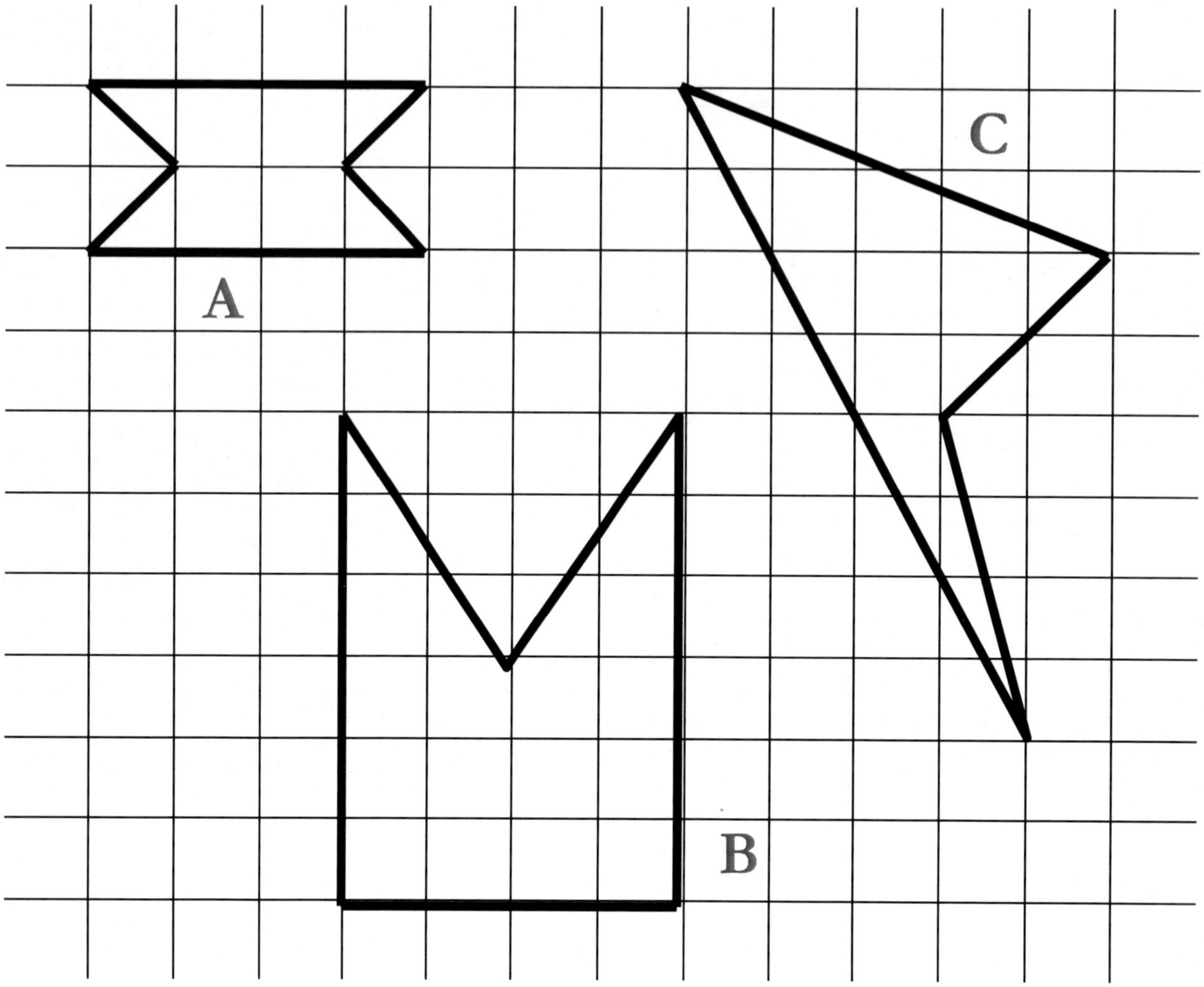

Write the letter of the shape which **does not have a line of symmetry**.

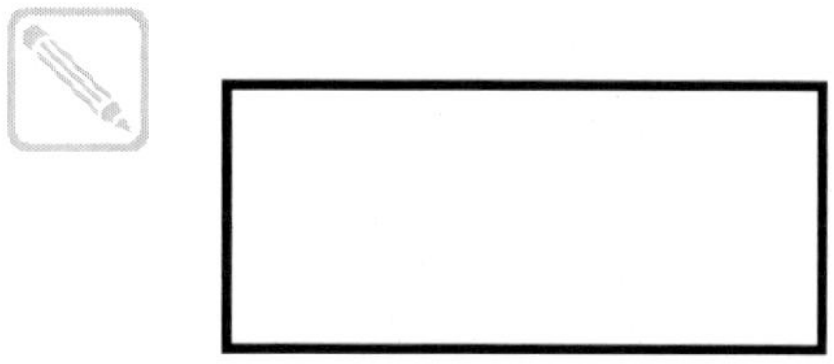

1 mark

13. Write the **missing numbers** in the boxes.

a). (4 x 2) + 87 – 21 = ☐

b). ☐ -- 24 + 135 = 261

2 marks

14. Which of these numbers **gives 80 when rounded to the nearest 10**.

Circle the correct answer.

55 29 82 47 88

1 mark

15. S and T each stand for a whole number.

S - T = 160

S is **three times bigger than** T.

Calculate the numbers for **S and T.**

Show your working

S =

T =

2 marks

16. All **260 students** are going on a school trip.

There are **28 seats** on the coach.

How many coaches are needed?

2 marks

17. If you **put 25 into this function machine, what will come out?**

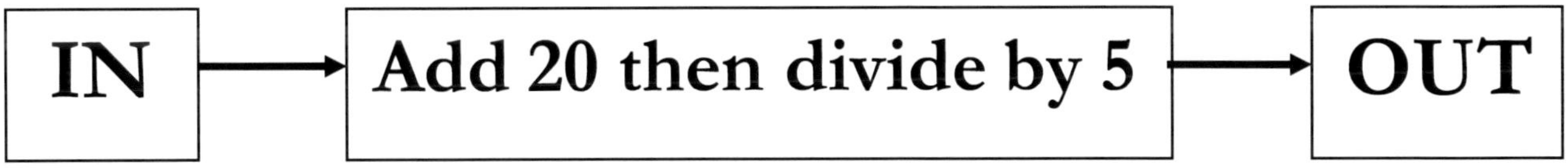

1 mark

18. Ben and Lisa a going to watch a film.

The film begins at **10 minutes to nine**.

Complete the clock below to show what time the film will start.

2 marks

19. Fill in the blanks.

☐ x 30 = 150

1 mark

20. A farmer has **250 sheep** on his farm. At the market he **sells 50** of his sheep and bought **14 pigs.**

How many animals does the farmer have on his farm **after he returns** from the market?

Show your working

2 marks

21. School lunch costs **£2.50 per day.**

Gemma buys school lunch for **4 days of the week only**.

How much money does she **spend** on school lunch in **one week**?

£

1 mark

22. There are **3 classes** in each year from 1 to 6. Each class has **21 children.** How many children are there in the **whole school**?

Show your working

2 marks

23. If today is **Monday 3rd of January**. What will be the date **next Monday**?

1 mark

24. Jenny invited **12 friends** to her birthday party. Her mother wanted to cut the cake evenly so that each person including Jenny will get **2 slices** of cake. How many slices of cake would she need to cut the cake into so that each person gets 2 slices?

2 marks

[END OF TEST]

Please do not write on this page.

Mathematics Test
Key Stage 2

Paper 13

Calculator NOT allowed

For marker's use only

Page	Marks
Total	

Instructions

You **must not** use a calculator to answer any questions in this test.

Work as quickly and as carefully as you can.

You have **45 minutes** for this test.

If you cannot do one of the questions, **go on to the next one.**

You can come back to it later, if you have time.

If you finish before the end, **go back and check your work.**

Follow the instructions for each question carefully.

This shows where you need to put the answer.

If you need to do working out, you can use any space on a page.

Some questions have an answer box like this:

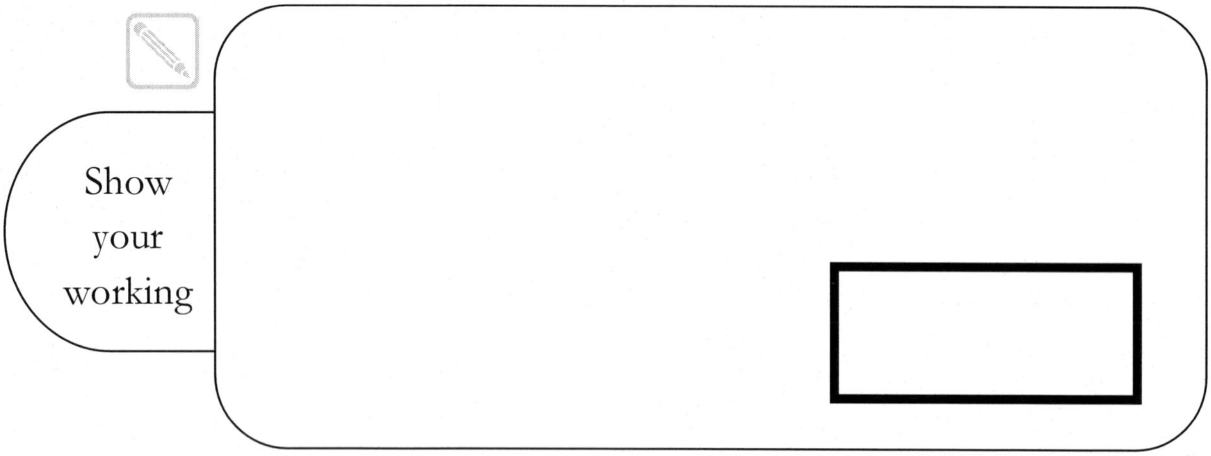

For these questions you may get a mark for showing your working.

1. **5 kg** of apples cost **£1.20.** How much does **8 kg** cost?

Show your working

£

1 mark

2. Arrange in **order of size, starting with the smallest.**

0.26 **0.82** **0.11** **0.52**

1 mark

3. Liam is **6 years older** than Tom. Their ages **add up to 78 years.**

How old is Tom?

Show your working

Years

2 marks

4. Draw lines **using a ruler** to show how this shape can be **divided into 4 triangles exactly.**

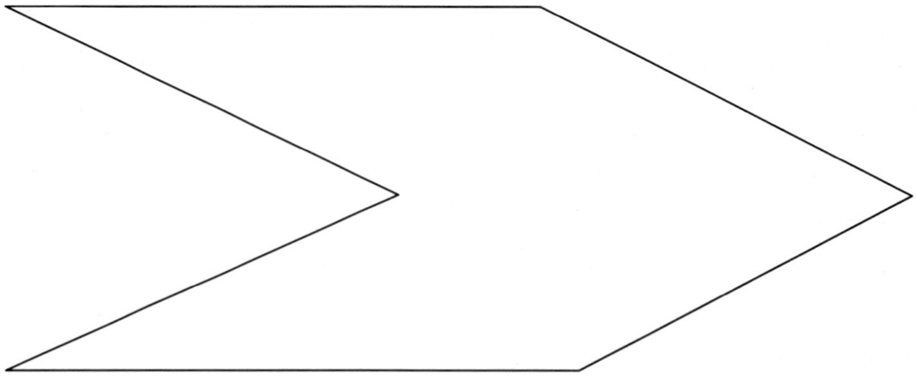

1 mark

5. Fill in the **missing numbers**.

$$\begin{array}{cccc} & \square & 3 & 2 \\ - & 6 & 2 & \square \\ \hline & 3 & \square & 1 \\ \hline \end{array}$$

2 marks

6. During a spring sale Amy bought a ladder for **£24**.

It was on sale with **40% off the original price**.

a). What was the **original price** of the ladder?

£

2 marks

b). How much did she **save**?

£

1 mark

7. The table below shows the number of each drinks sold on Christmas morning at a cafe.

Type of drink	Number sold
Cappuccino	20
Latte	18
Hot chocolate	9
Mocha	3
Tea	5

a). Draw a **bar chart** to show this information.

2 marks

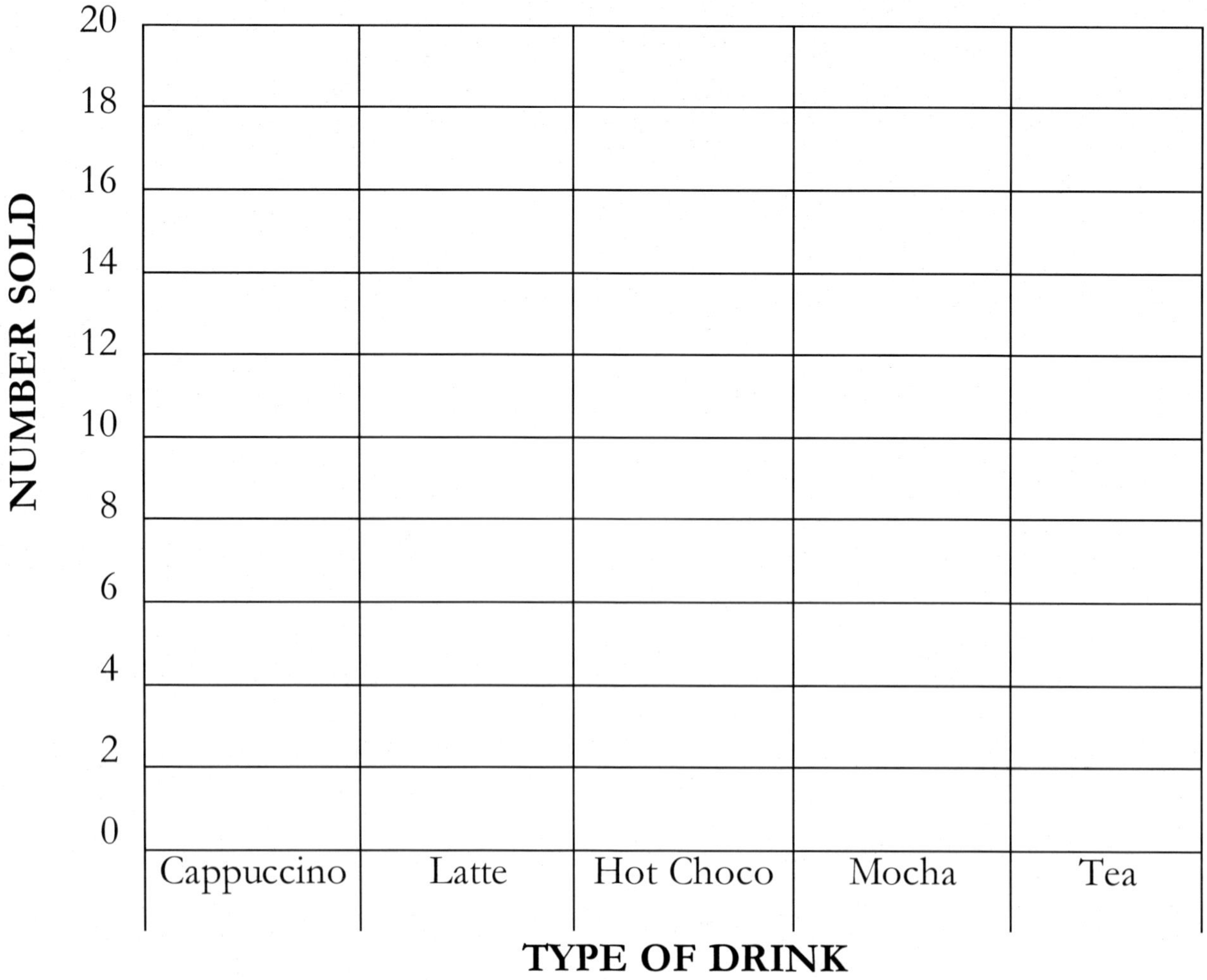

b). **All together** how many hot chocolates, mochas and cappuccinos were **sold**?

1 mark

8. Calculate **850 – 123**

1 mark

9. Calculate **412 ÷ 2**

1 mark

10. A muffin costs **15p more than** a cookie.

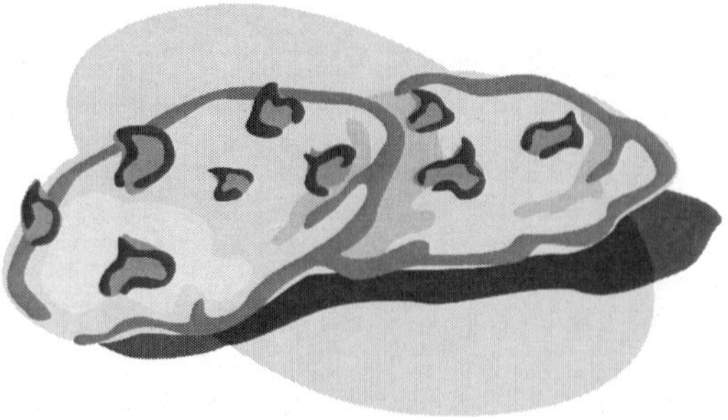

Joanna bought **a muffin and two cookies for 90p**.

How much does a muffin and cookie cost.

Show your working

Muffin = **p**

Cookie = **p**

2 marks

11. Write in the **missing numbers**.

Number	Rounded to the nearest **whole** number
6.05	
6.55	
5.45	
5.54	

2 marks

12. Mia and Lisa are both washing cars.

Mia gets £23 and Lisa gets £37.

They share what they get **equally** between them.

How much does **each** of them get?

£

2 marks

13. The table below shows the cost of a new swimming kit.

Item	Cost
Swimming costume	**£ 10.50**
Goggles	**£ 5.25**
Hat	**£ 3.20**

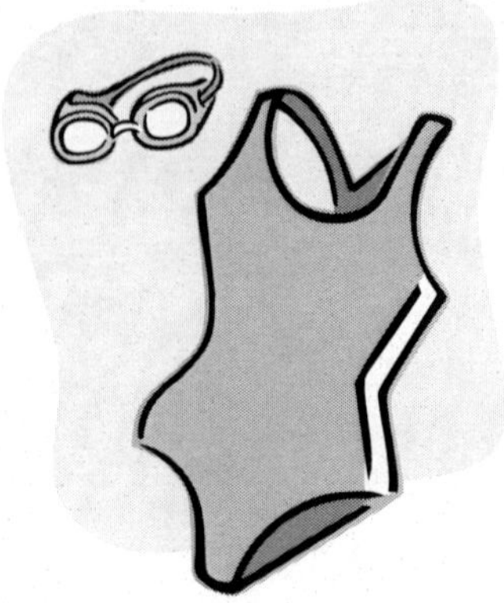

Altogether , how much does a **complete swimming kit cost**?

£ ______

2 marks

14. Write in the **missing digits** to make this correct.

	☐	7	☐
+	4	5	6
	7	2	8

1 mark

15. Calculate **6.23 + 4.77**

1 mark

16. What is **809** rounded to the **nearest hundred**?

1 mark

17. Estimate the sum by **rounding each number to the nearest thousand** and then add.

2277 + 3598

1 mark

18. Nine people go to the cinema together

Adults £9.90 per movie Children £ 4.90 per movie

There are **three adults** and **six children**.

a) How much does it cost for the **adults in total** for one movie?

£

1 mark

b) How much does it cost for the **children in total** for one movie?

£

1 mark

c) How much does it cost **altogether** for one movie?

£

1 mark

Drinks cost **60p** each.

d) **How many** people had a drink if they spent £4.80?

1 mark

19. What is the area of the **shaded** region? Both the shapes are squares.

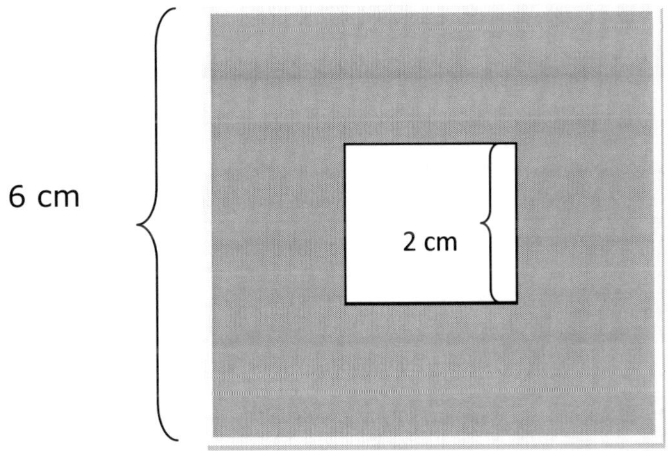

Show your working

cm^2

2 marks

20. Alex had £ 9.50. He spent £ 2.25 on sweets.

Alex gave his two friends £ 1.20 each.

How much money was **left**?

Show your working

£

2 marks

21. A car travels at a speed of 60 miles per hour.

How far will it travel **in 4 hours**?

miles

1 mark

22. A motorbike is traveling **60 kilometers per hour.**

How many meters does the motorbike travel in **one minute**?

m

1 mark

23. If Mike had lunch at £20.50 and he gave 20% tip, **how much did he spend**?

Show your working

£

2 marks

24. Andy gave half of his stamps to John.

John gave half of his stamps to James.

James gave 1/4 of the stamps given to him to Peter and kept the remaining 6.

How many stamps did Andy start with?

Show your working

2 marks

[END OF TEST]

Please do not write on this page.

Mathematics
Key Stage 2

Paper 14

Calculator NOT allowed

For marker's use only

Page	Marks
Total	

Instructions

You **must not** use a calculator to answer any questions in this test.

Work as quickly and as carefully as you can.

You have **45 minutes** for this test.

If you cannot do one of the questions, **go on to the next one.**

You can come back to it later, if you have time.

If you finish before the end, **go back and check your work.**

Follow the instructions for each question carefully.

This shows where you need to put the answer.

If you need to do working out, you can use any space on a page.

Some questions have an answer box like this:

For these questions you may get a mark for showing your working.

1. In a zoo, the adult elephant weighs 3 times more than a crocodile.

An adult elephant

A crocodile

Together their weight is **700 kilograms.**

How much does the adult elephant weigh?

Show your working

kg

2 marks

2. **Complete** the diagram below to make a shape that is symmetrical about the mirror line.

Use a ruler. You may also use a mirror or tracing paper.

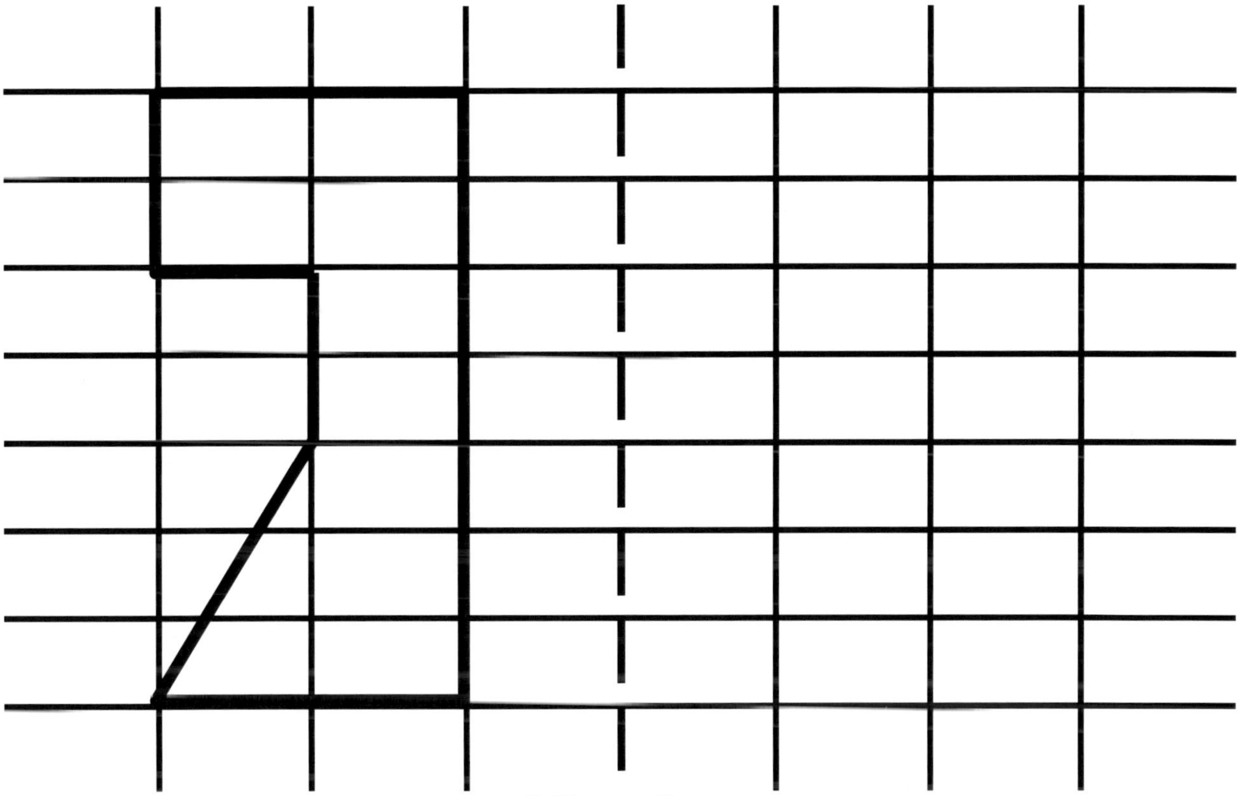

2 marks

3. Calculate **360 - 50 + 26**

1 mark

4. What number is **marked at A.**

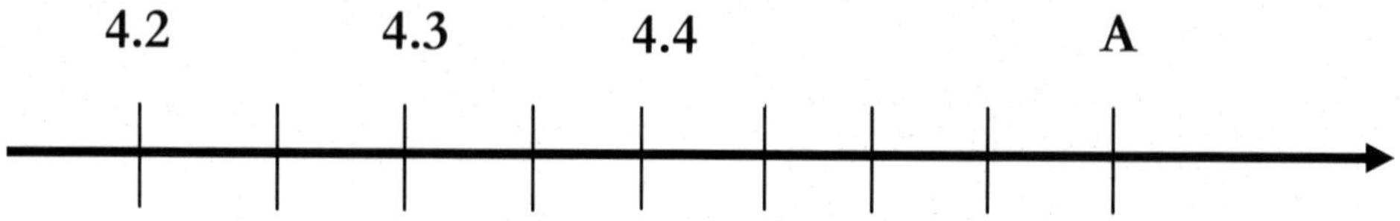

1 mark

5. The pictogram shows the number of ice-creams Lisa sold at her shop in a day.

 = stands for **10 ice-creams**

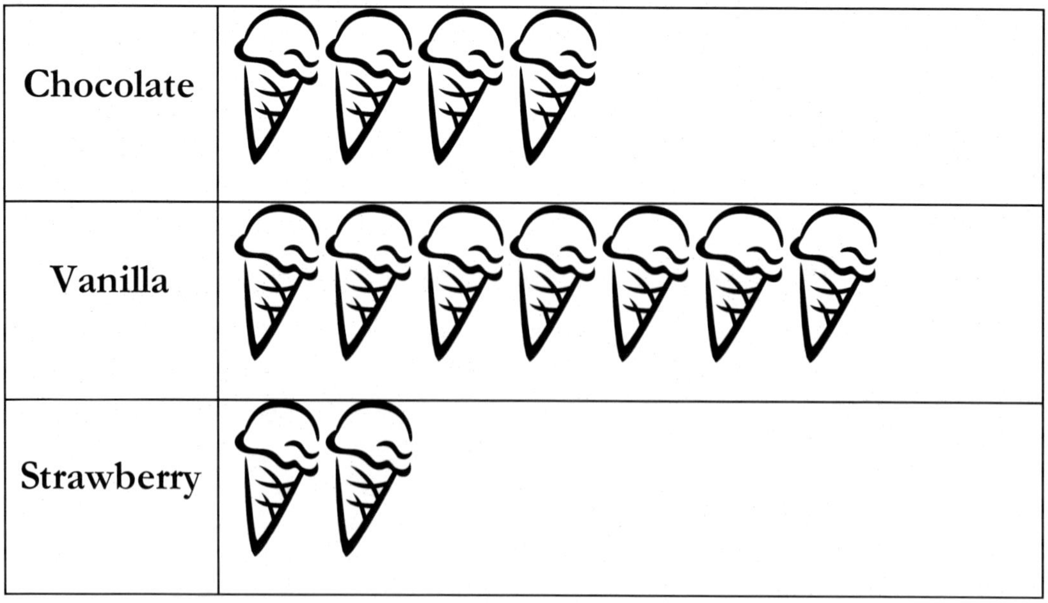

a). How many ice-creams were **sold altogether**?

1 mark

b). How many **more** vanilla than strawberry ice-creams were sold?

1 mark

6. At a local book shop the books were on sale for 50p.
Liam wants to buy 14 books.
He only has £5.50.
How much **more** money does Liam need?

Show your working

£

2 marks

7. Five school football teams took part in the summer club football match.

Each team played six matches.

The table below shows how many matches each team won, drew or lost.

	Won	Drew	Lost
King Henry's School	1	3	2
Timmins	2	2	2
St Catherine's School	3	3	0
Harpur School	4	0	2
Newham School	1	4	1

a). How many teams **drew more** matches than they lost?

1 mark

b). Each teams scores two points for a win,

one point for a draw,

no points for a loss.

Which team scored the most points?

1 mark

8. List the following amounts of money in **order from smallest to largest.**

£2.66	£1.81	£2.81	£2.69	£2.24

£	£	£	£	£

2 marks

9. Which sign makes the statement true? Choose from the following <, > or =.

0.8 + 7.9 ☐ **9.7**

1 mark

10. Marion's dog eats 0.3 kilograms of dog food each day.
How much dog food will Marion's dog eat in **5 days?**

☐ **kg**

1 mark

11. The normal price of a lamp is £ 15. You can choose one of two special offers in a sale. What would you pay if you choose the **best** offer in the sale?

25 % off

1/3 off

What would you pay if you choose the **best** offer in the sale?

2 marks

12. **Circle** the number that is closest to 400.

338 **406** **4040** **388** **398**

1 mark

13. The number **12** goes into two of the square of this multiplication grid.

Tick (✓) the two squares where the 12 goes.

X	1	2	3	4	5
1					
2					
3					
4					
5					

2 marks

14. Calculate **214 x 6**

1 mark

15. Tick (✓) each box that is exactly ¼ **shaded.**

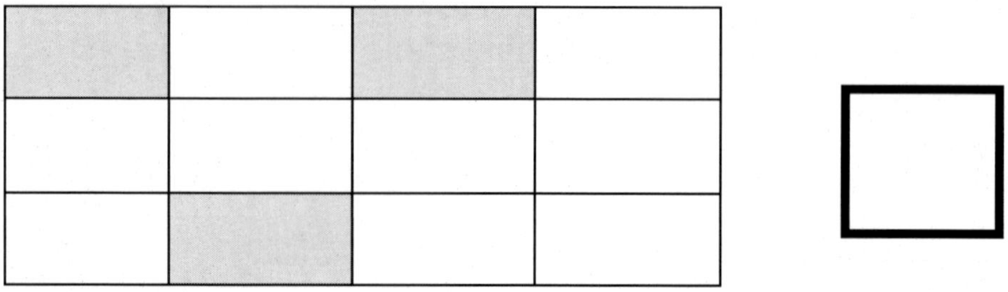

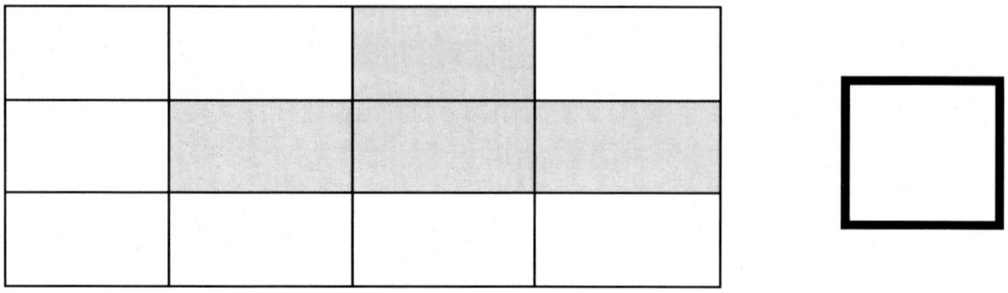

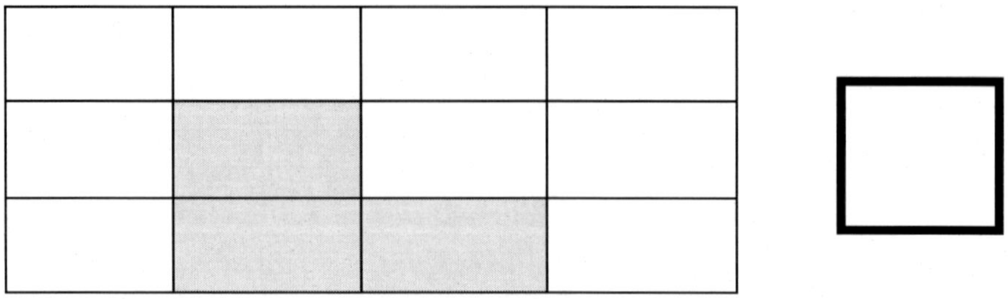

2 marks

16. Lisa receives £5 pocket money each week. If she saves half of it for 10 weeks, how much will she have saved?

£

2 marks

17. A can of coke is sold for 30p. How much **saving** is there on each can when 6 cans are sold for £1.50?

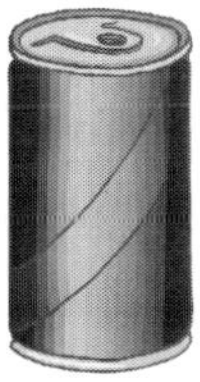

Show your working

p

2 marks

18. Liam won £ 2050 in the lottery and gave away £ 450.
How much does he have **left**?

£

1 mark

19. On the grid below, **mark clearly with a dotted line** the line of symmetry.

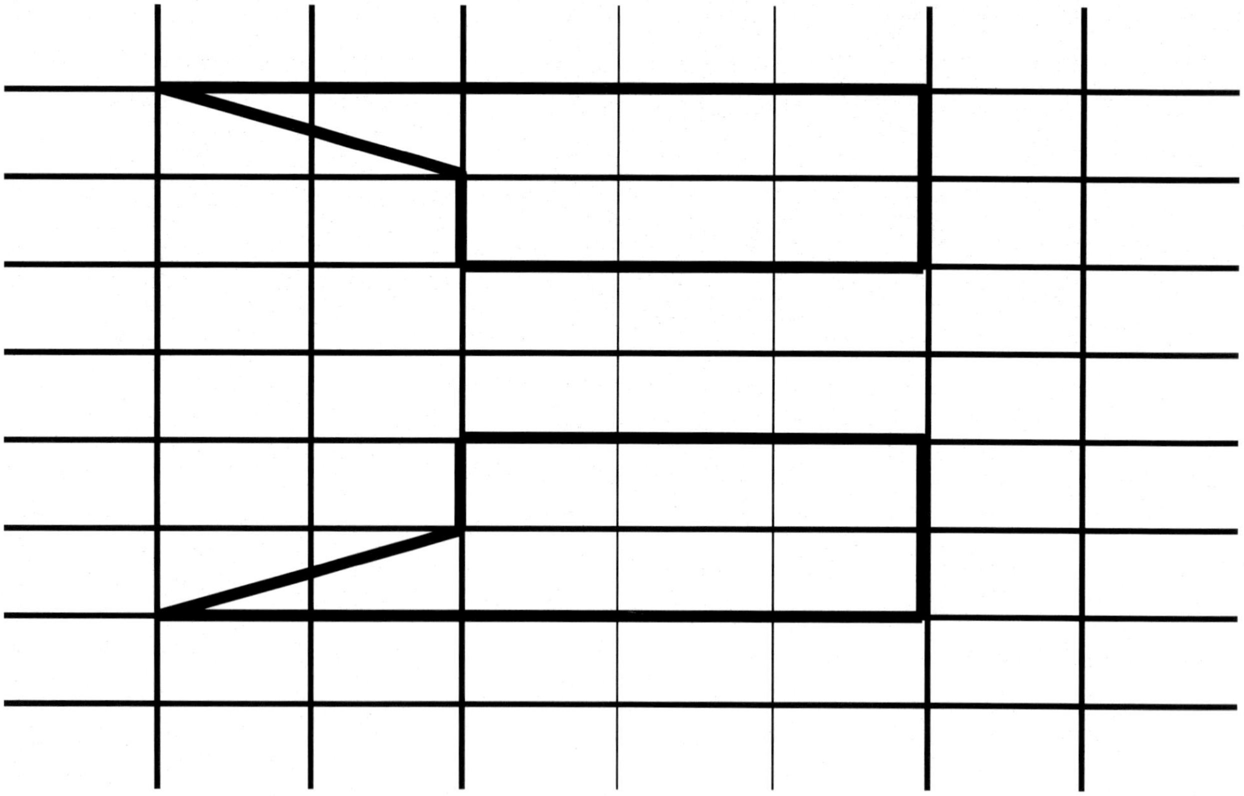

2 marks

20. What decimal number is **coloured** ?

2 marks

21. Andy earns £10 per hour. Last week, he worked 3 hours on Monday, 4 hours on Tuesday, and 5 hours on Wednesday. He had Thursday off, and then he worked 8 hours on Friday. How much money in total did Andy **earn** last week?

Show your working

£

2 marks

22. Estimate the difference. Round each number to the nearest whole number, then subtract.

$$18\ ^{2}/_{5} \quad - \quad 15\ ^{3}/_{5}$$

The difference is approximately

2 marks

23.

$$^{3}/_{9} + ^{2}/_{6} + ^{1}/_{3} =$$

1 mark

24. Subtract:

4 hours 25 minutes – 1 hours 30 minutes =

hours minutes

2 marks

[END OF TEST]

Please do not write on this page.

Mathematics
Key Stage 2

Paper 15

Calculator NOT allowed

For marker's use only

Page	Marks
Total	

Instructions

You **must not** use a calculator to answer any questions in this test.

Work as quickly and as carefully as you can.

You have **45 minutes** for this test.

If you cannot do one of the questions, **go on to the next one.**

You can come back to it later, if you have time.

If you finish before the end, **go back and check your work.**

Follow the instructions for each question carefully.

This shows where you need to put the answer.

If you need to do working out, you can use any space on a page.

Some questions have an answer box like this:

For these questions you may get a mark for showing your working.

1. What is the **lowest common** multiple of 3 and 11?

1 mark

2. The pictogram below shows the number of cars that came to the car wash in a week.

 = stands for 5 cars

Day of the week	Number of cars that came to the car wash
Monday	
Tuesday	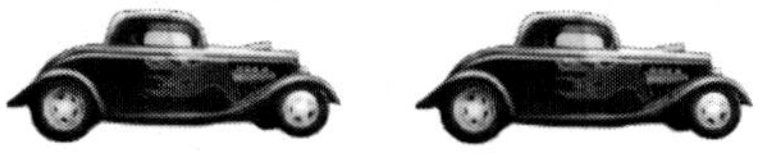
Wednesday	
Thursday	
Friday	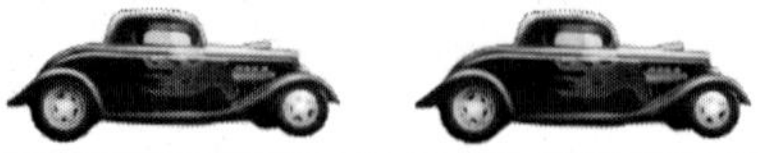
Saturday	
Sunday	

a). How many cars came to the car wash **on Thursday**?

1 mark

b). How many cars came **altogether for the whole week**?

2. marks

c). What **percentage** of the cars came to the car wash on Sunday compared to that whole weekend?

Show your working

%

2 marks

3. Write these in order starting with the smallest.

25p **£20** **£1.20** **5p** **£150**

1 mark

4. Calculate **535 x 7**

1 mark

5. Calculate **248 / 4**

1 mark

6. Liam has five number cards shown below.

5	2	18	6	3

Use two of these cards to make the calculation below correct.
One has been filled out for you

$\square \times 6 + \square = 23$

2 marks

7. Write the missing numbers in the boxes.

a). $\square$ **x 28 = 168**

b). **426 −** $\square$ **= 374**

c). $\square$ **x 2 = 96**

2 marks

8. There are **14 crayons in a box**.

The school **buys 21 boxes** of crayons.

How many crayons does the school buy?

Show your working

2 marks

9. a). On this grid below plot the points **A (1 , 2), B (2 , 8), C (4 , 8)** and **D (3 , 2).**

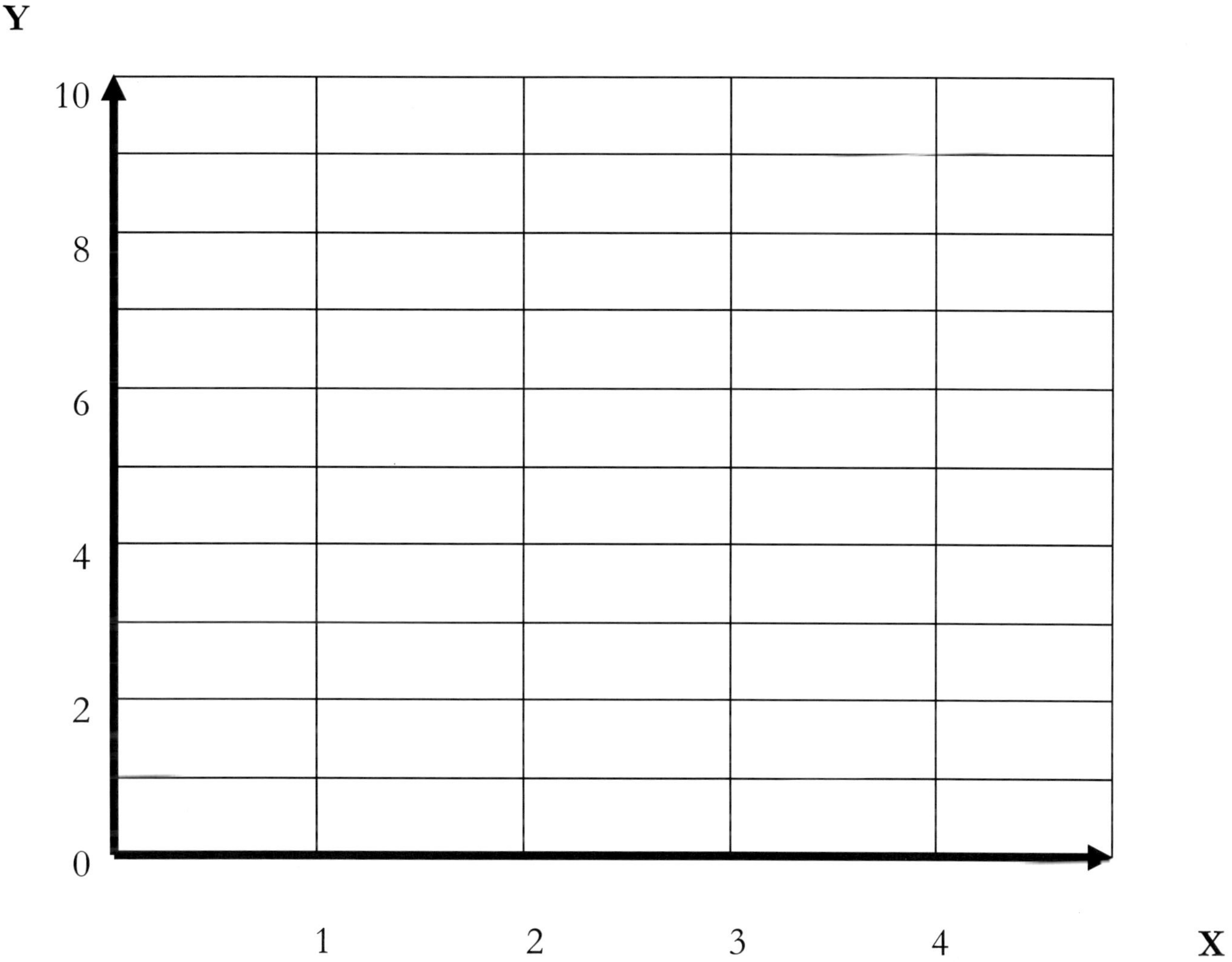

2 marks

b). Name the shape:

____

1 mark

10. Here is a clock:

a). What time was it **50 minutes ago**?

1 mark

Here is another clock:

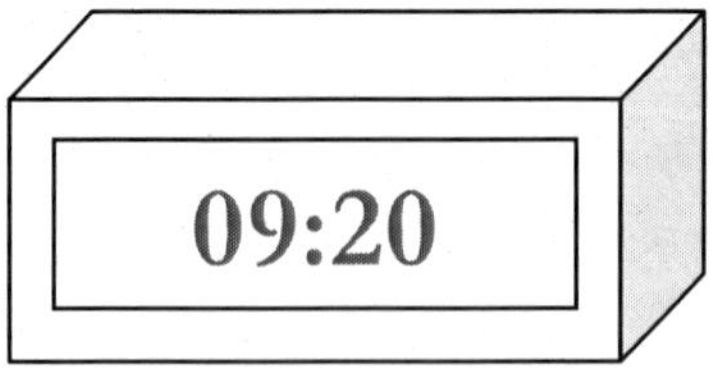

b). What time is it in **35 minutes time**?

1 mark

11. Write each of the following numbers in the correct place on the Venn diagram shown below.

20 **12** **53** **25**

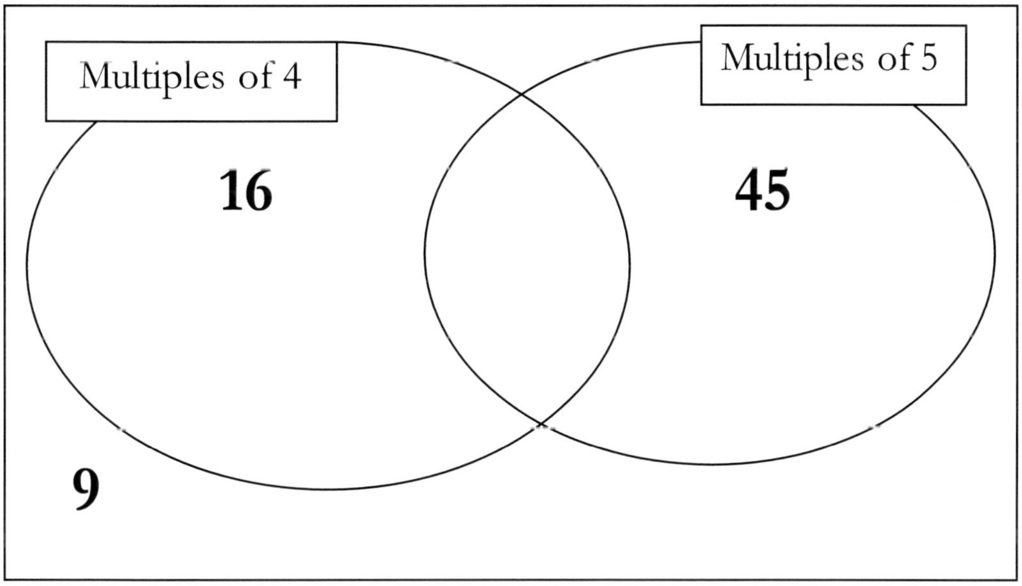

2 marks

12. Multiply

$$\begin{array}{r} 3\ \ 2\ \ 1 \\ \times \qquad\quad 4 \\ \hline \\ \hline \end{array}$$

1 mark

13. Write in the **missing time**?

10:15 10:45 11:15 [ ] 12:15

1 mark

14. In 217, in **which place is the 2**?

	Tick (✓)
Tens	
Hundreds	
Thousands	

1 mark

15. Put these numbers in **order from the smallest to the largest.**

64.5 **0.4** **5.4**

1 mark

16. Work out **5/6 - 1/3 =**

1 mark

17. Select the pairs of decimals which **add to 10.**

a 5.3 + 5.7

b 3.7 + 7.3

c 7.3 + 2.7

1 mark

18. Andy is sending three parcels weighing 350g, 600g and 470g.
What is the **total weight** of the three parcels?

Show your working

g

1 mark

19. A tree grows 20 cm every year. It was 6.2 metres tall five years ago.
How tall is the tree now?

Show your working

m

2 marks

20. Five bricks have a total weight of 1,500g.
What is the **weight of three bricks**?

g

2 marks

21. There are 6 more girls than boys in a class. If there are 32 children in total in the class, **how many boys** are there?

2 marks

22.

$$2x - 1 = 7$$

x =

1 mark

23. Which **two numbers** total 600?

219, 334, 275, 486, 381

and

2 marks

24. Hannah buys a new dress for £ 26.50. She pays for it with £ 30.
How much **change** will she get?

£

1 mark

25. A computer game has three levels of difficulty. On the first level, 2000 points can be scored. On the last two levels, 4000 points per level can be scored. What is the **total score** for all three levels?

1 mark

[END OF TEST]

Please do not write on this page.

Mathematics
Key Stage 2

Paper 16

Calculator NOT allowed

For marker's use only

Page	Marks
Total	

Instructions

You **must not** use a calculator to answer any questions in this test.

Work as quickly and as carefully as you can.

You have **45 minutes** for this test.

If you cannot do one of the questions, **go on to the next one.**

You can come back to it later, if you have time.

If you finish before the end, **go back and check your work.**

Follow the instructions for each question carefully.

This shows where you need to put the answer.

If you need to do working out, you can use any space on a page.

Some questions have an answer box like this:

For these questions you may get a mark for showing your working.

1. Write the **missing number** in the box.

8 x ☐ **= 280**

1 mark

2. Calculate **692 – 155**

☐

1 mark

3. Complete these fractions so that they are equivalent to ½.

___	___	___	___
2	16	8	24

2 marks

4. The below shows the increase in bus fare.

Bus fares	
Old fare	New fare
£ 2.70	£ 4.20
£ 1.50	£ 3.00
£ 5.60	£ 7.80
£ 2.10	£ 2.90
£5.20	£6.50

a). Mia's new fare is £ 7.80.

How much has her **fare increased** by?

£

1 mark

b). Liam's new fare is £ 1.30 more than his old fare.

What is his **new fare**?

£

1 mark

5.

		4	☐	5
+		☐	4	☐
	1	1	8	9

2 marks

6. Circle the numbers given below, find the numbers which are **divisible by 2 without a remainder.**

15 21 26 36 39

2 marks

7. Round the numbers to the nearest 10. One has been done for you.

3321 **3320**

2589 ☐

4075 ☐

836 ☐

2 marks

8. Place the correct **symbol, >, = or <** in the box given below.

a). 965 [ ] 857

b). 88 [ ] 99

c). $\frac{12}{2}$ [ ] 8 – 2

2 marks

9. Write the **missing number** in the calculation below.

2 . 1 X [ ] = 420

1 mark

10. What is the fraction of square is **not shaded**?

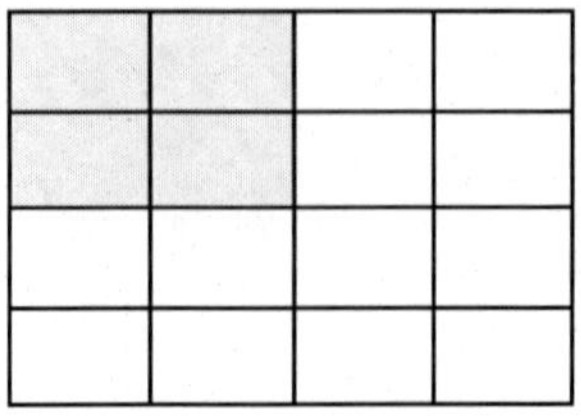

2 marks

11. Calculate **25% of £500**

£

2 marks

12. Calculate **11.84 – 9.95 =**

1 mark

13. **x** stands for a whole number.

$6x - 10 = 32$

What is the value of X =

2 marks

14. **Match** each shape to the correct information.
One has been done for you.

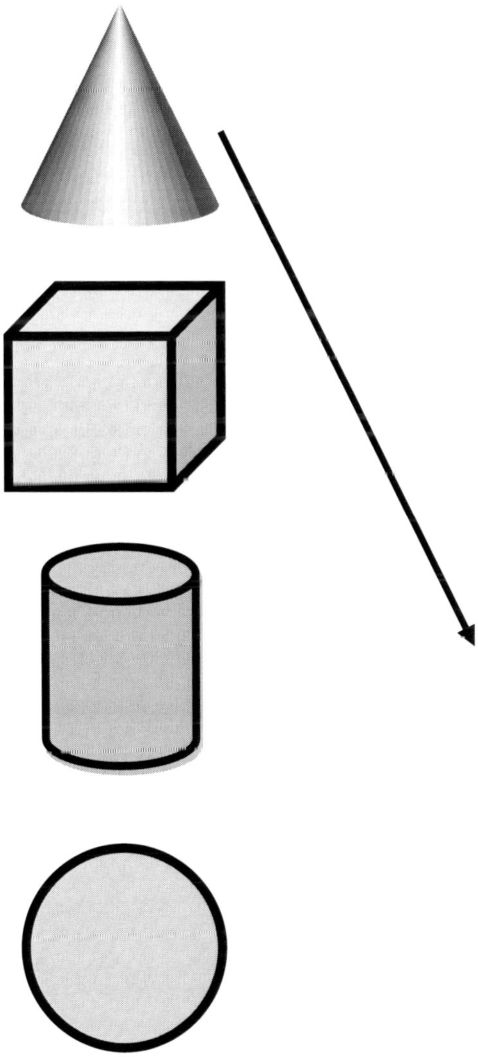

I have 1 face which is curved.

I have 3 faces.
2 faces are plane and 1 is curved.

I have 2 faces.
1 is plane and the other is curved.

I have 6 faces.
All are equal in shape and size.

2 marks

15. Year 6 children go on a trip to a theme park.

3 teachers also go along with them.

	Ticket Cost
Adults	**£7.50**
Children	**£5.00**

There are 30 children in Year 6.

What is the **total cost** for the whole group to enter the theme park?

Show your working

£

2 marks

16. Liam has 3 cartoons of milk. Each carton contains 150 ml of milk.

Liam shares the cartoons of milk between 6 children.

How much milk will they each get?

[ ] ml

2 marks

17. Fill in the blanks.

a). 22 x 10 = [ ]

b). 25 x [ ] = 2500

c). 5 x [ ] = 8 x 5

d). (3 + 4) x 10 = (3 x10) + [ ]

2 marks

18. Nine people go to pizza hut together.

The meal cost £40.50.

They all decide to split the cost evenly.

How much does **one person** pay?

£

1 mark

19. A number sequence has the rule:

Take 4 away from the previous number, then add 12.

Complete the number sequence using this rule.

21 | | |

2 marks

20. Here is a digital clock.

14:15

What time was it **25 minutes ago**?

2 marks

21. **Circle** the number that is closest to 650.

125 662 525 595 432

1 mark

22. Calculate **5% of 360**

1 mark

23. Look at the four number cards.

8	1	4	2

Choose three cards to make an **odd number less than 300.**

1 mark

24. Write each of the following in the correct place on the diagram below.

21 48 24

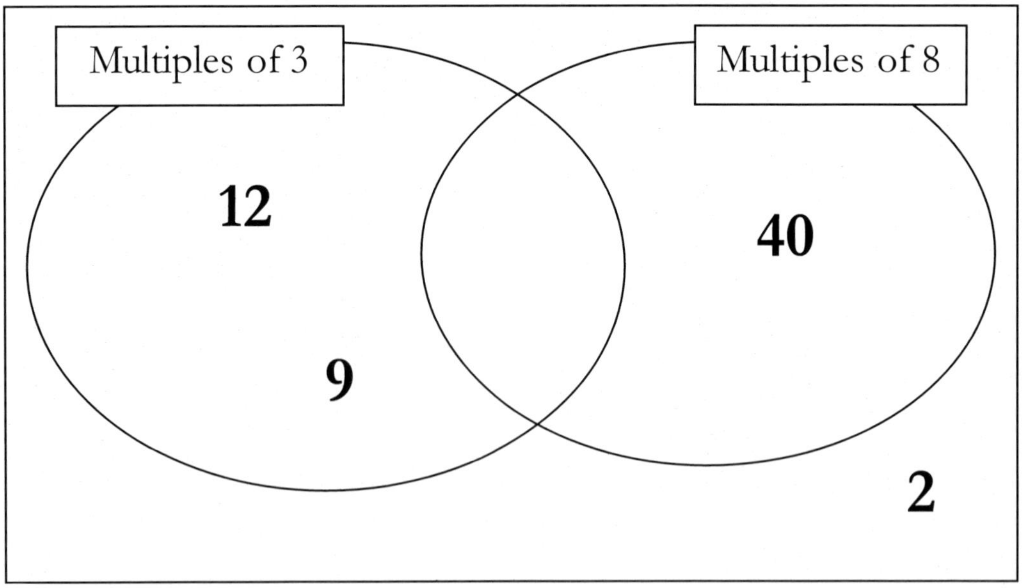

2 marks

[END OF TEST]

Please do not write on this page.

DETAILED ANSWERS
WITH
COMPLETE MARK SCHEME

Mathematics Paper 1

1. Calculate.

14896 - 5348 = 9548

56 x 39 = 2184

2. What is the 5/8 of 128? 80

128/8 = 16

16 x 5 = 80

What is the 1¼ of 128?

128 + 32 = 160

(128/8) x 2 = 32

3. Divide 2896 by 8.

362

4. Multiply 58 and 96

58 x 96 = 5568

5. Please find the perimeter of the below shape.

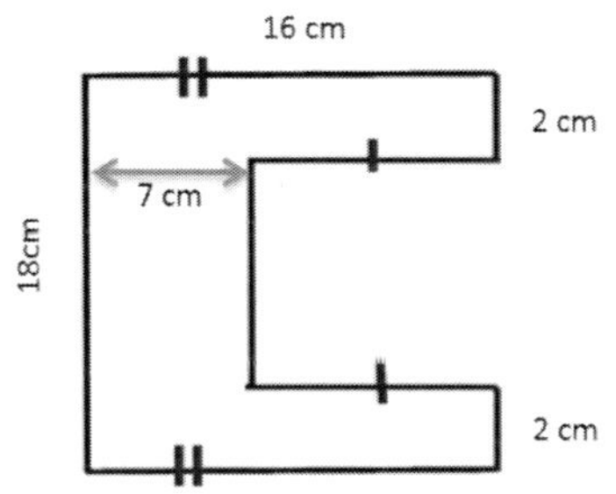

(16cm x2) + (18cm x2) + (9cm x2)

= 32cm + 36cm + 18cm

=86cm

6. Following are the prices of groceries in a store.

Item	Price
Apple	78p / 100grams
Milk	88p / 1 litre
Cake	£1.5/ 500grams
Chocolate	90p/ 200grams

Sara buys the following.

- Apple 500grams
- Milk 500 millilitre
- Cake 750grams
- Chocolate 500grams

What is the total cost of the items Sara purchased?

(0.78/100 x 500) + (0.88 x 2) + (1.5/ 500 x 750) + (0.9/200 x 500)

3.9 + 1.74 + 2.25 + 2.25 = £10.16

If she gives a £20 note, how much balance will she get?

£20 - £8.84 = £11.16

7. Which number in between 50 and 60 can be divided exactly by both 7 and 8? 56

Because 7x8 = 56

8. Below are the details of three children.

Jack	Peter	Joanne
32.75Kg	**38.23Kg**	**29.98Kg**

What is the total weight of all three children?

32.75 + 38.23 + 29.98 = 100.96kg

How much heavier is Peter than Joanne?

38.23 – 29.98 = 8.25kg

If Jack reduces 1kg, what is the mean weight of all three children?

99.96/3 = 33.32kg

9. Calculate.

X= 6

$X^2 + (3 + 5x) – 4x =$

$6^2 + (3+ 5x\ 6) – 4x\ 6$

= 36 + (3+30) – 24

= 36 + 33 – 24

= 45

10. Add the below numbers together.

- Two hundred and sixty eight
- Three thousand and ninety two
- Nine thousand six hundred and six

268 + 3092 + 9606 = 12,966

11. Fill the gaps.

9586 - [6629] = 2957

9586 - 2957 = 6629

[5418] - 3986 = 1432

3986 + 1432 = 5418

12. Amy is 9 years and 3 months old. Her brother John is 2 years and 7 months elder to Amy. How old is John? 11 years 10 months

9 years 3 months + 2 years & 7 months = 11 years & 10 months

13. Last year a school bought 189 new chairs and removed 53 old chairs. A charity donated them 236 chairs. At the end of the year they had 1985 chairs. How many chairs did they have in the beginning of last year?

1613

1985 – (189 – 53 + 236)

= 1985 – 372

= 1613 Chairs

14. Joanne has only 2 pence and 5 pence coins in her till. Altogether she has 30 coins. She counts all the coins and says that she has £1.20. She also says the number of 5 pence is double the amount of 2 pence. How many 2 pence and 5 pence she had in the till?

Number of 2 pence = x

Number of 5 pence = 2x

(x * 2 pence) + (2x * 5 pence) = £1.20

2x + 10 x = 1.20

12x = 1.20

x = 10

So,

Number of 2 pence = 10

Number of 5 pence = 20

15. In a bag there 12 green balls, 5 red balls and 3 blue balls.

If John closes his eyes and picks one ball, what is the probability of getting a red ball?

Total number of balls = 12 + 5 +3 = 20

Probability of getting red ball = 5/20

Sara takes out two green balls. Now what is the probability of John getting a red ball?

Total number of balls = 10+5+3 = 18

Probability of getting red ball = 5/18

16. Calculate.

15 x (28 -4) ÷ 2 + 25

Use BODMAS

15 x 24 ÷ 2 + 25

= 15 x 12 + 25

= 180 + 25

= 205

17. Sara gets 17 questions right in a test out of 40 questions. What percentage of the questions did she get right?

(100/ 40) x 17 = 42.5%

18. If a square's perimeter is 64cm, what is the area?

Square's one side length = 64/4 = 16cm

Area = 16 x 16 = 256cm²

19. If the time is 9.17p.m, what is the time in 24 hour clock?

12 + 9.17 = 21.17

20. Three elephants weigh 1 ton 500grams. What is the total weight in grams?

1 ton 500grams = 1000.5 kg

1000.5 x 1000 = 1000500grams

Because – 1000grams = 1 kg

1000kg = 1 ton

21. What is the next two numbers?

1, 5, 9, 13, 17, 21

5-1 = 4

9-5 = 4

13-9 = 4

Therefore, 13+ 4 = 17

17+4 = 21

22. Find b.

12a + 15b = 294

6a – 2 = 70

a is equals to = 6a – 2 = 70

6a = 70 + 2

6a = 72

a = 72/6 = 12

Therefore,

12a + 15b = 294

12x 12 + 15b = 294

144 + 15b = 294

15b = 294 – 144

15b = 150

b = 10

23. What is the difference between 1.25 metre and 278 millimetres?

1.25meter = 1.25 x 100 = 125cm

125cm = 125 x 10 = 1250 millimetres

1250 – 278 = 972 millimetres

Because – 10 millimetres = 1 cm

100 cm = 1 metre

24. How many eights are there in 1264?

158

1264/8 = 158

Mathematics Paper 2

1. Put the following in order of size, starting from the smallest.

¼ 2/5 6/8 ½

Common denominator for all four fractions is 40.

Therefore,
¼ = 10/40
2/5 = 16/40
6/8 = 30/40
½ = 20/40

The order as follow,
1/4 , 2/5, ½, 6/8

2. If two numbers multiply 35 and their total is 12 when adding, what are the two numbers?

5 & 7
Because,
5x7 = 35
5+7 = 12

3. Amy takes 36 minutes to reach her school. What time does she have to leave home if she has to reach school at quarter past eight?

7.39 am.

4. Fill in the blanks.
Example - The first 4 multiples of 3 add ups to 45
(3+6+9+12+15= 45)

The first 3 multiples of [7] add ups to 42
7+14+21 = 42

The first [5] multiples of 4

add ups to 60

4+8+12+16+20 = 60

5. Shade 2/3 of this pattern

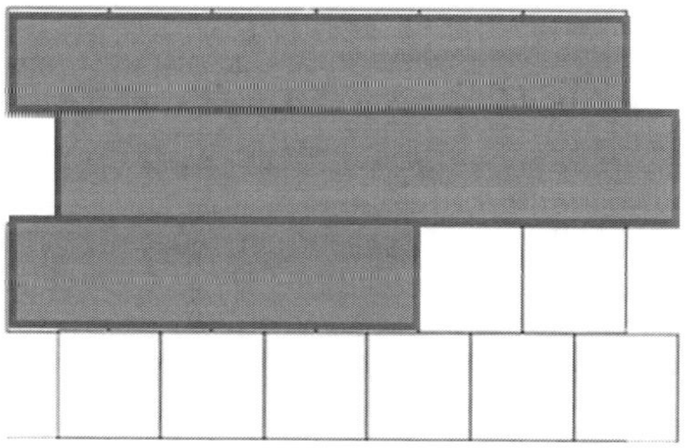

24 squares

Therefore, 24/ 3 x 2 = 16 squares should be shaded

6. The table shows the weight of four children

Name	Weight
Tina	28kg
Peter	34kg
Ben	38kg
Rita	36kg

Amy says the average weight of all four children is equal to Peter's weight.
Is Amy correct? Yes
Explain why?

Total weight = 28+34+38+36 = 136kg

Average weight = 136/4 = 34kg

7. Jacob stays in a hotel for 9 days. He pays £530; this includes the hotel charges for 9 days and an additional charge for a bottle of water which cost £8. How much does that hotel charge for one day?

9 days hotel charges = £530 – 8 = £522
Hotel chargers for one day = £522/ 9 = £58

8. Rita played in the park from 5.15pm to 7.10 pm. How long did she spend in the park?
1 hour and 55 minutes

5.15 to 6.15 = 1 hour

6.15 to 7.10 = 55 minutes

9. The diagram shows the favourite fruit of 90 children in a class.

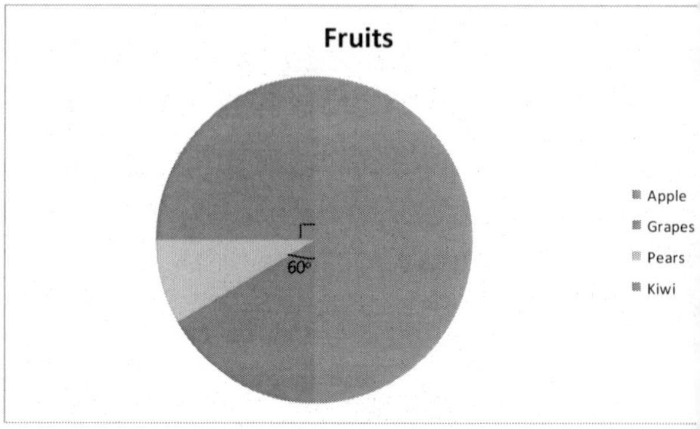

- What angle has been used to indicate pears?

  90 – 60 = 30°

- How many children in the class like grapes?

  360° = 90 children

  Grapes = 60°

  60° = (90/360) x 60 = 15 children

10. The second hand of the clock points at number 2. After 90 seconds what number will it be pointing to?
8

Again it will come to 2 at 60 seconds.
Number 2 = 10 seconds
Therefore, 10 + 30 = 40 seconds
40 seconds = number 8

11. The time in USA is 5hours behind UK. Rita wants to be the first to call her friend in USA for her birthday at USA time 12 am. At what time does she have to call her at the UK time?
5 am

Because UK is 5 hours forward from USA time

12. There are few number cards as shown below.

9	5	8	3	2

- Using all these cards once only
  - Make a largest number

9	8	5	3	2

  - Make the smallest odd number

2	3	5	8	9

13. Amy had marbles in a bag.
To play a game with 5 groups, she gave 7 marbles to each group of children. Therefore, one marble was left off in the bag.
5*7 +1 = 36
For another game with 7 groups, she gave 5 marbles to each group and one marble was left off.
7*5+1=36
For the final game with 6 groups, she gave 6 marbles for each group and had no marble left.
6*6 = 36
How many marbles did she have in the bag?
36

14. Two numbers in a ratio 2:5.
One number is 30.
There are two possible values for the other number. What are the two possible values?
75 & 12

If 30 – 2
5 = 30/2 *5 = 75

If 30 = 5
2 = 30/5 * 2 = 12

15. In a class room, children were asked what their favourite ice cream was. The result is shown in the below Venn diagram.

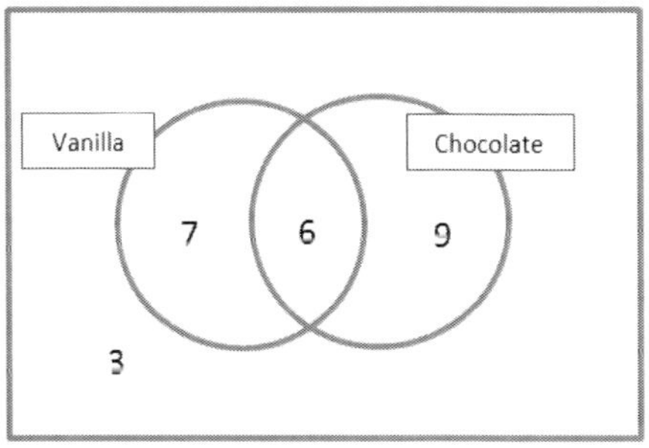

What percentage of the class room like both vanilla and chocolate ice cream? 24%

Total number of children = 3+7+6+9 = 25

Number of children who like both vanilla and chocolate = 6

Answer in percentage – (100/25) *6= 24%

What percentage of the class room does not like both vanilla and chocolate ice cream? 12%

Number of children who do not like both vanilla and chocolate ice cream = 3

In percentage = (100/25)*3= 12%

What percentage of the class room does not vanilla ice cream? 48%

Number of children who do not like vanilla ice cream = 9+3 = 12

In percentage = (100/25)*12 = 48%

16. How many seconds a baby has lived when he/she is one year old?
Number of seconds in one day = 60*60*24 = 86400 seconds
One year = 86400 * 365 = 31536000 seconds

17. There is a cuboid tank full of water. The measurement of the tank is as below.

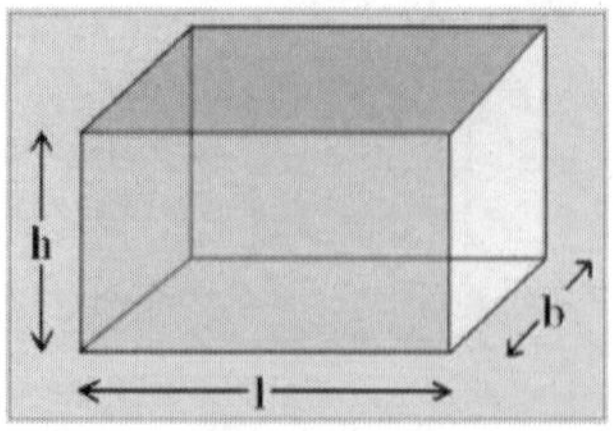

h	10cm
l	20cm
b	15cm

Amy tries to remove all the water from the tank by using a bucket. The bucket can contain 150cm^3 of water at a time. If she fills the bucket fully, how many times does she have to use the bucket to clear all the water in the tank? 20 times

Volume of the tank = h x l x b = 10 x 20 x 15 = 3000cm^3
Number of times to use the bucket = 3000/ 150 = 20 times

18. A box of chocolates cost £65. There are 25 chocolate bars in the box. Each bar contains 5 pieces of chocolate. How much does a piece of chocolate cost? £0.52

Cost of a chocolate bar = 65/25 = £2.60

Cost of one piece chocolate = 2.6/5 = £0.52

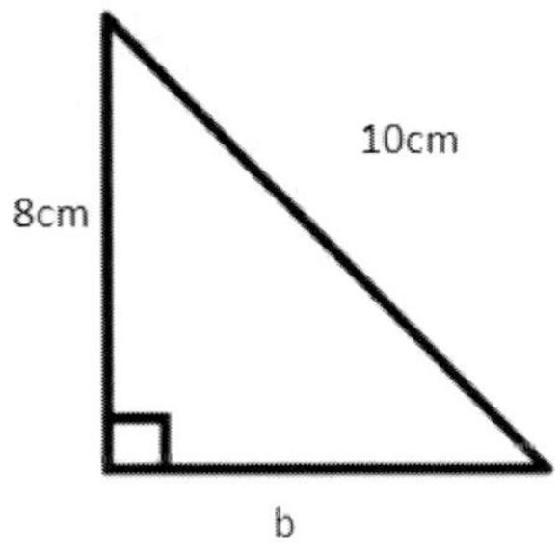

19. Find b.

Use Pythagorean Theorem.
$8^2 + b^2 = 10^2$
$64 + b^2 = 100$
b^2 – 100-64 =36
b = 6cm

20. What fraction of a hour is 20 minutes?

20/60 = 2/6= 1/3

21. A bottle can contain 5.25 litres. It is 70% filled with water. How many more water is need to fill the bottle fully? Provide the answer in millilitre. 1575ml
100 – 70 = 30% not filled
30% = (5.25/100) * 30 = 1.5751
In millilitre = 1.575 * 1000 = 1575ml

22. 8 boxes of chocolate cost £10. How much does 10 boxes of chocolate cost? £12.5
One box of chocolates' cost = £10/8 * 10
= £1.25 * 10
= £12.50

23. Find B.

6 B
X B

621

B = 9
Because in the below table only 9*9 ends with 1

2*2	4
3*3	9
4*4	16
5*5	25
6*6	36
7*7	49
8*8	64
9*9	81

24. John prints his project report. The total cost for it is £5.65. Below are the prices for printing at the store.

Printing (per page)	25p
Binding	£1.05
Cover	85p

How many pages did John print? 15

Cost for printed pages = Total cost – (cover cost + binding cost)
= £5.65 – (£0.85 + £1.05)
=£3.75
Number of pages printed = £3.75 / £0.25 = 15 pages

Mathematics Paper 3

1. Choose a number from each set to get the answer.

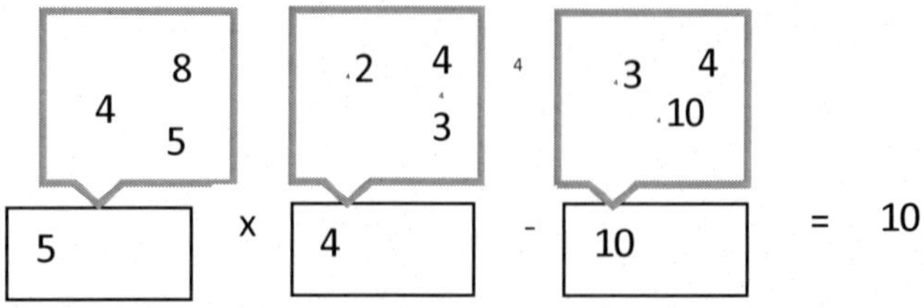

2. Lucy, Aaron and Sharon went to the garden to pluck fruits.
The below table shows the number of fruits they collected.

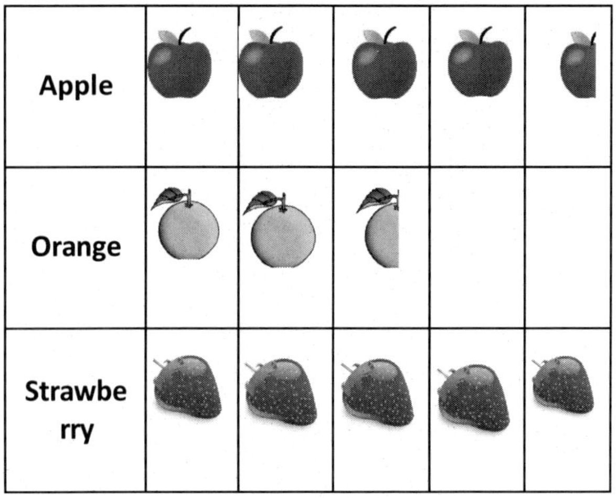

Key	
	Represents 10 apples
	Represents 10 oranges
	Represents 10 strawberries

How many apples have they plucked more than oranges?

Total number of apples = 45

Total number of oranges = 25

Difference = 45 – 25 = 20

All together how many fruits have they plucked?

Total number of apples = 45

Total number of oranges = 25

Total number of strawberries = 50

All together = 45 + 25+ 50 = 120

If they sell each strawberry for 40p, how much will they gain?

£0.40 * 50 = £20

3. Aaron buys a book for £1.60. He gives three coins to the seller and gets 40p as balance.
What are the three coins he would have given?

£1 and two 50pence

4. Sharon makes a 3D shape using 2 triangles and 3 rectangles. What shape is Sharon making?
Triangular Prism

Draw it.

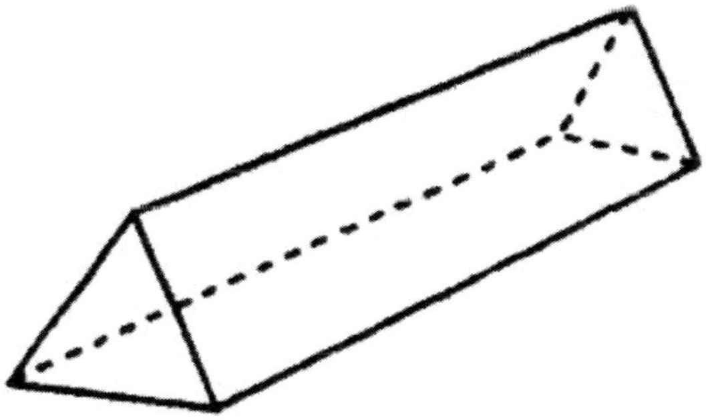

5. Calculate.
190 – (11-8)2 x 12 =
Use BODMAS

190 – 9 X 12
=190 – 108
=108

6. Aaron has x number of candies in a box. He equally divides it among 12 friends. Each person gets 9 candies. How many candies were there in the box?

12*9 =108

7. Sharon goes to school in her bicycle. She rides in a speed of 5km/hour. Her school is 2km from her home. If she leaves home at 7.45am at what time will she arrive at school?
8.09 am

5km = 60 minutes
2 km = 60/ 5 * 2
= 12*2
= 24 minutes

Therefore, 7.45 + 24 = 8.09 am

8. Fill the blanks.

8	+	9	=	17
9	÷	6	=	1.5
8	x	7	=	56

17-8 = 9

9/6 = 1.5

56/ 7= 8

9. Calculate
568 x 1.6
= 908.8

10. Complete the number sequence

10, 25, 40, 55, 70
25-10= 15
40-25= 15

Therefore, 40+15 =55
55+15= 70

38, 30, 22, 14, 6
38 – 30 = 8
30 – 22 = 8

Therefore, 22 – 8 = 14
14 – 8 = 6

11. Draw the mirror image.

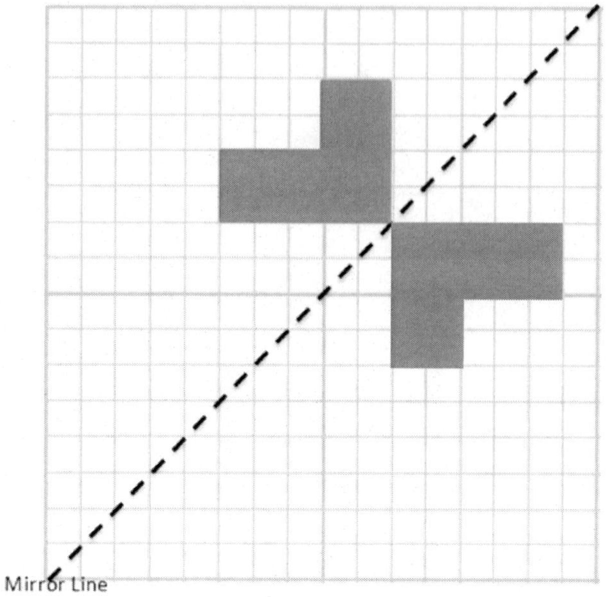

12. Calculate.
9/15 ÷ 27/35
= 9/15 x 35/27
= 9*35/ 15*27
= 315/ 405
=63/81
=7/9

13. Write the correct sign =, < and > in the blank

14	-	7	<	18	-	9
12	x	5	=	180	÷	3
15	+	8	>	19	+	3
30	-	16	<	14	+	9

14. Lucy and Sharon each buy a meal. Lucy pays £10 and gets a balance of £1.25. Sharon's meal cost £3.65 more than Lucy's meal.

What is the price of Sharon's meal?

Lucy's meal price = £10 -£1.25
=£8.75

Sharon's meal price= £8.75 +£3.65
= £12.40

15. What is the volume of the below object?

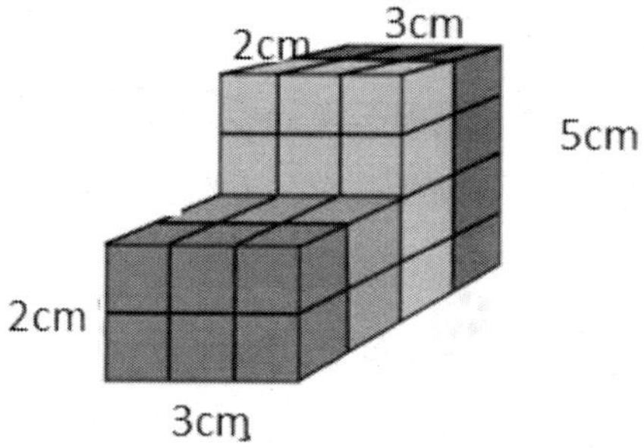

Bigger rectangular prism volume + smaller rectangular prism volume

= (5*2*3) + (2*3*3)
=30 + 18
48cm^3

16. 10 apples weighs 1.5kg. Aaron buys 8 apples. What is the weight of the apples Aaron bought?
10 apples = 1.5kg
8 apples = 1.5/10 * 8
= 1.2kg

If 1kg apple cost £2.5, how much will Aaron pay for 8 apples?

1 kg = £2.5

1.2kg = 2.5/1 * 1.2

= £3

17. Calculate.

19586 ÷ 8

= 2448.25

152 x 12

= 1824

18. Add 0.182 + 1.82 + 18.2+ 182

= 202.202

19. Below is a sketch of a shape (not to scale).

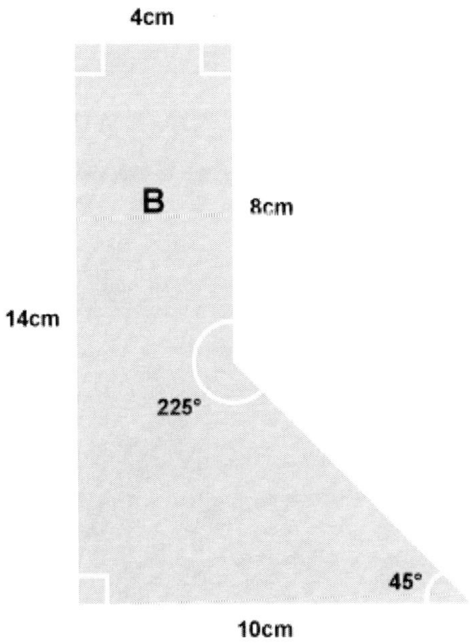

Use angle measurer and ruler and draw the shape accurately.

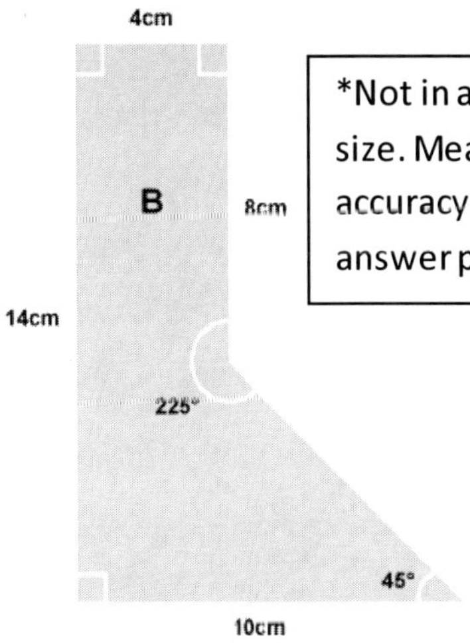

*Not in accurate size. Measure the accuracy in the answer paper.

20. Lucy has a standard dice. When she spins it, what is the probability of getting an odd number?

Odd numbers in dice- 1,3 and 5
Therefore, probability of getting an odd number is 3/6

21. Write the prices in order, starting from the smallest.

£6.96	£6.89	£0.99	£6.99

£0.99, £6.89, £6.96, £6.99

What is the difference between the highest price and the lowest price?

£6.99 - £0.99 = £6

22. Calculate.

398 – 259.78 = 138.22

23. Sharon bought two apples and two oranges.
Apple cost 20p more than orange.

The total cost was £1.20. What is the price of one apple and one orange?

Price of orange = x
Price of apple = x + 20

Therefore,

x +x+ x+20 +x +20 = 120p
4x + 40 = 120
4x = 120 – 40
4x = 80p
x= 20p

Orange price = 20p
Apple price = 40p

24. Three different numbers add up to 30. All three numbers are even and less than 15. What are the three different numbers could be?

Possible answers

4, 12, 4

4, 10, 6

2, 10, 8

Mathematics Paper 4

1. Join the following equations to the nearest number

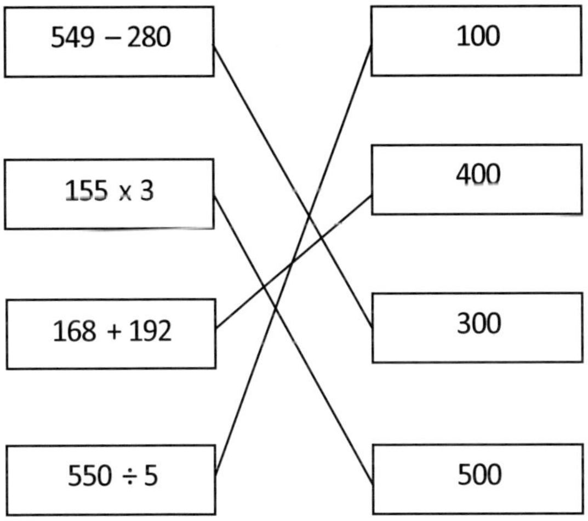

2. Write the missing digits.

5	5	+	9	5	=	150

Since the answer's last digit is zero and the second numbers last digit is 5, the first number's last digit should be 5. Because, 5+5=10

Therefore the second number will be, 150 – 55 = 95

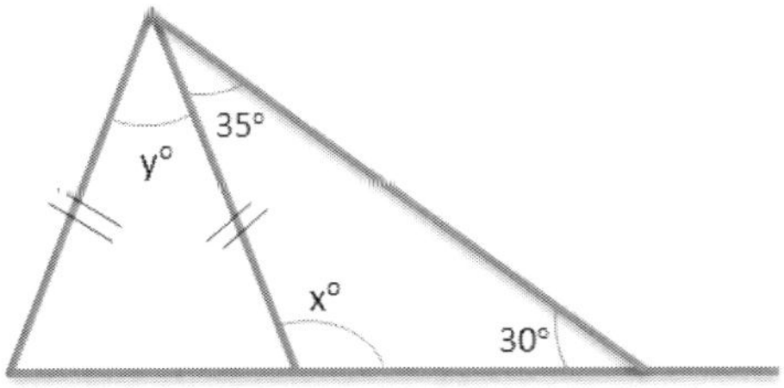

3. Find x and y.

The sum of the degree measures of angles in any triangle equals 180 degrees.

Therefore,

X° = 180 – (35+30)

X° = 180 – 65

X° = 115°

Degree of angle of a straight line = 180°

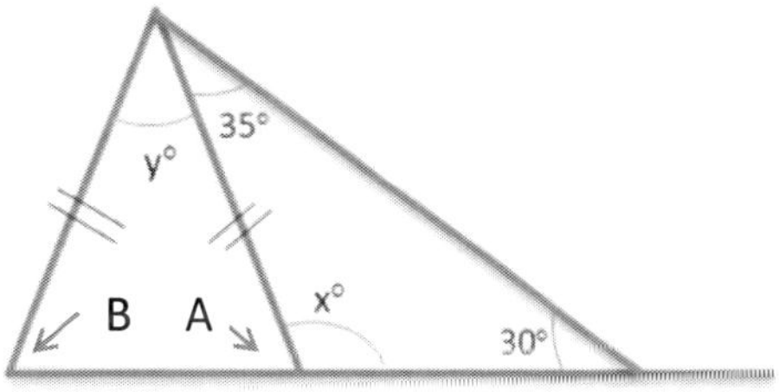

Therefore A = 180 – 115 = 65°

When two sides are equal the angles are also equal.

Therefore, A = B

Therefore, y = 180 – (A+B)

= 180 – (65+65)

=180 – 130

=50°

4. The time is 11.25am. Lucy says the time is closer to 11am than 12pm.

Explain why? A hour has 60 minutes. The median is 30 minutes. Since 11.25 is in the first 30 minutes Lucy says it is closer to 11 am than 12 pm.

If it is passed 11.30 it will be closer to 12.

5. Cross all the diagrams which are ¼ shaded

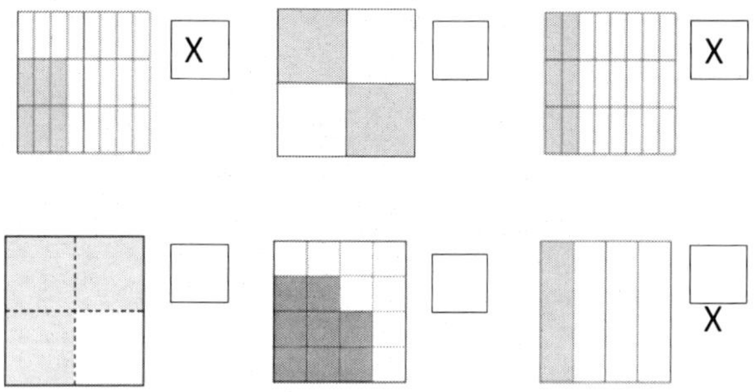

6. There are 40 children in a class. Each child needs 3 strawberries each. There are 20 strawberries in a box. How many boxes of strawberries do they need for all 40 children?

Total number of strawberries needed = 40 * 3 = 120

Number of boxes is = 120 / 20 = 6 Boxes of strawberries are needed

7. Which is not a quadrilateral ?

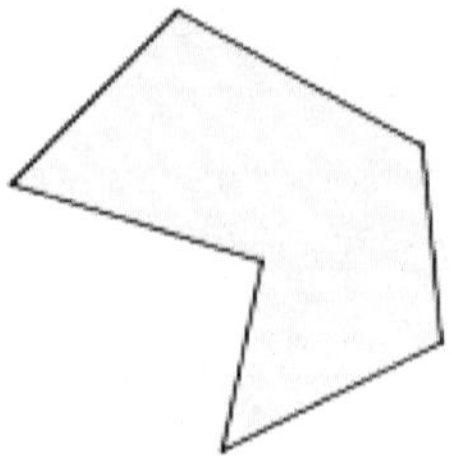

A quadrilateral is a polygon with four sides (or edges) and four vertices or corners.

8. Write the below numbers in an order, starting from the smallest.

1.86   8.16   18.6   0.186

0.186, 1.86, 8.16, 18.6

9. In a class test, Aaron finished in 18 minutes. Sharon finishes 5 minutes before Aaron. Lucy finishes 9 minutes after Sharon.

How long does Lucy take to finish the test?

Time taken by Sharon = 18 – 5 = 13 minutes

Time taken by Lucy = 13 + 9 = 22 minutes

If the test has started at 10.15am, at what time did Sharon finish?

10.15 am + 13 minutes = 10.28am

10. Sharon buys a 250ml bottle of water. She drinks 70% of the water. How much water did Sharon drink? Provide your answer in ml.

70% = (250/100) *70
= 175ml

11. Sharon and Lucy each have a till. Altogether they have £25. Sharon puts £2.5 to Lucy's till, so that they both have same amount of money.

How much money did each have in the start?
Sharon – 12.5 + 2.5 = £15
Lucy – 12.5 – 2.5 = £10

Now each has – 25/2 = £12.5

12. There is 30 marbles in a bag. Aaron takes 15 marbles. From the balance Sharon takes 60% of the marbles.

Who has the highest number of marbles?
Aaron

Explain why?
Aaron has – 15 marbles
Sharon has = (30 – Aaron's marbles)/100 * 60
= 15/100 * 60
= 9 marbles

13. Calculate.
2.8 + (5.6 – 2.8) + 3.9 = 9.5

14. Sharon rotates the arrow 270° anticlockwise about the centre of the grid. Draw the arrow in its new position.

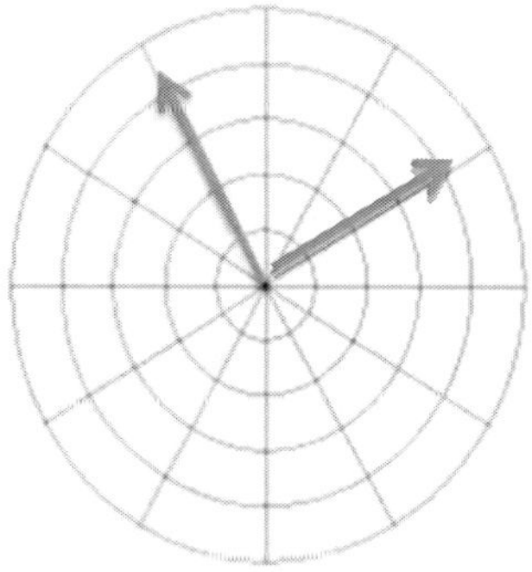

15. 1kg cake cost £2.80. 7 friends buy 9kg of cake and share the cost. How much does each person pay?

Cost of 9kg cake = 2.80 * 9 = £25.20
Each person pays – 25.2/7 = £3.60

16. Sharon has some rectangles and squares.

Using those she makes the below shape.

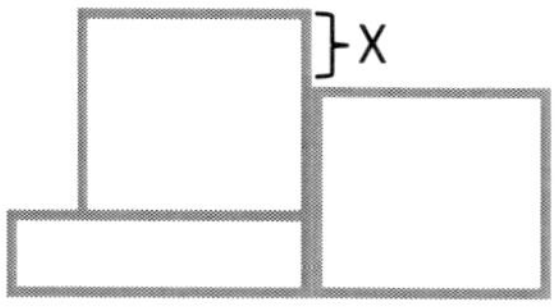

Find the length of X?

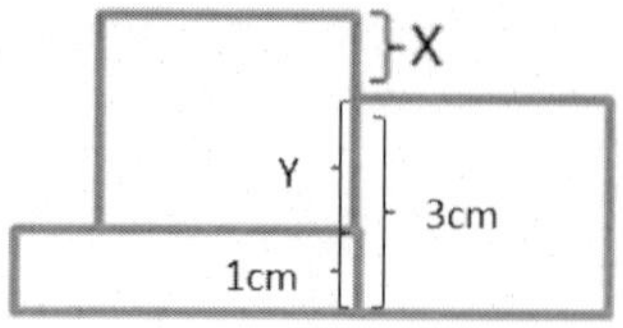

Y = 3 – 1 = 2cm

Therefore, x = length of the square – 2 = 3 – 2 = 1 cm

What is the area of the above image?

Area of 1 square = 3*3 = $9cm^2$

Area of rectangle = 1 * 3 = $3cm^2$

Area of the image = area of 2 squares + area of a rectangle

= (9 * 2) + $3cm^2$

= 18 + 3

= $21cm^2$

17. Two whole numbers between 30 and 50 make 1710 when multiplied.

What are the two numbers?

45 and 38

18. Draw ten to seven in the below clock.

19. Aaron says "multiples of 10 ends in 0".

Is Aaron correct?

Yes

Explain why?

10 * 1 = 10

10*2 = 20

....

10*15 = 150

20. Write the missing number.

15 x a x 9 = 1620

15 x a = 1620/9

15 x a = 180

a = 180/ 15

a = 12

15 x 12 x 9 = 1620

26 + b - 16 = 29

26 + b = 29 + 16

26 + b = 45
b = 45 – 26
b = 19
26 + 19 – 16 = 29

21. Lucy buys 8 Roses and 5 Daffodils. Roses cost £1.50 more than daffodils.
Altogether it cost £31.50.

Write an expression for this.
If daffodils price = a
13a + (1.5*8) = 31.50
How much does single rose and daffodil cost?

Rose – £3 (Daffodil price + 1.50 = 1.50 +1.50 = 3)
Daffodil – £1.50

13a + (1.5*8) = 31.50
13a + 12 = 31.50
13a = 31.50 – 12
13a = 19.50
a = 19.50/13
a = 1.50

22. Sharon, Aaron and Lucy go to super market. Their bills totals are as follow.
Sharon - £29.65

Lucy - £35.25

Aaron - £40.65

Aaron rounds each bill total to full number and adds it.

What is the total amount Aaron gets?

30 + 35 + 41 = £106

Lucy adds up three bills exactly. What is the difference between Lucy's total and Aaron's total?
Lucy's total = 29.65 + 35.25 + 40.65 = 105.55

Difference is = 106 – 105.55 = £0.45

23. Write the smallest whole number to make the statement true

95 + 59 > 153
Because 95 + 59 = 154

24. Every 200g cake has 120g protein. A box of cake weighs 1kg and has 30 equal pieces. How much protein is there in on cake piece?

Protein in 1kg cake = 120/200 * 1000 = 120 * 5 = 600g

Protein in one cake piece = 600/ 30 = 20g

Mathematics Paper 5

1. Fill up the blanks.

11	23	35	47	59	71

47 – 35 = 12
59 – 47 = 12
Therefore, 11 + 12= 23
59 + 12 = 71

2. The numbers in the two circles add up to the number in square.

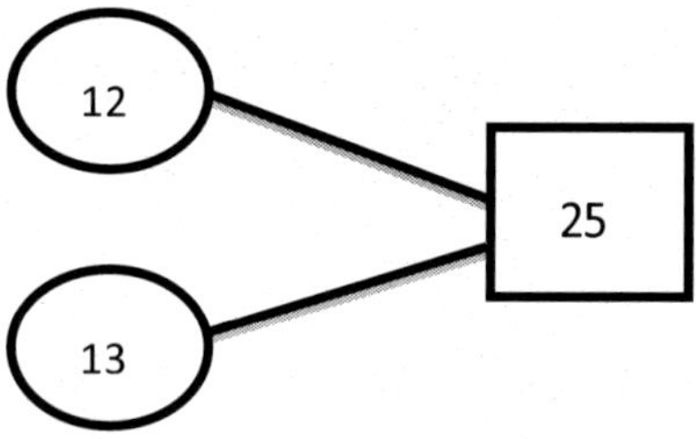

Using the same rule, write the missing numbers.

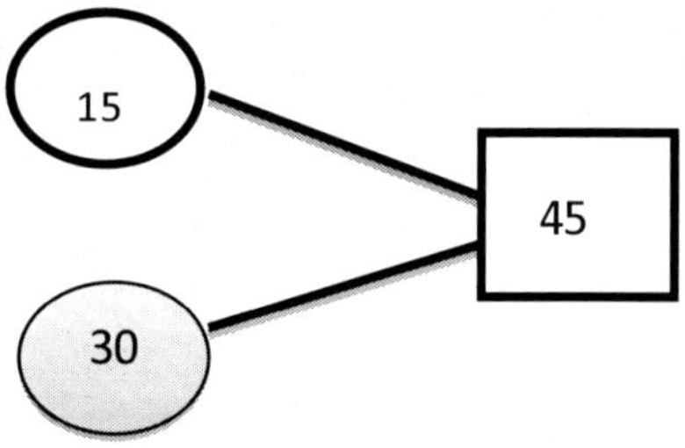

45 – 30 = 15

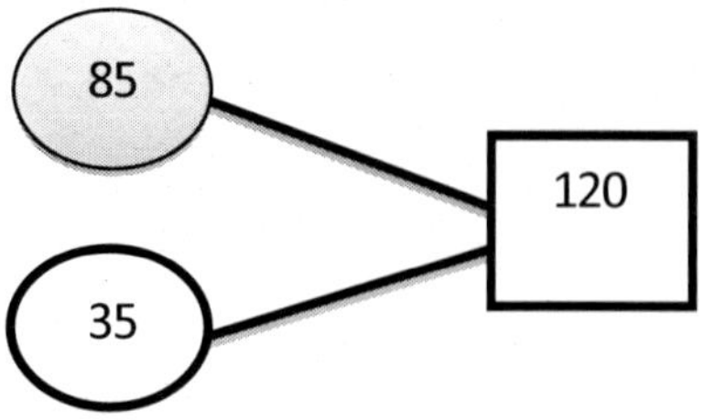

120 – 35 = 85

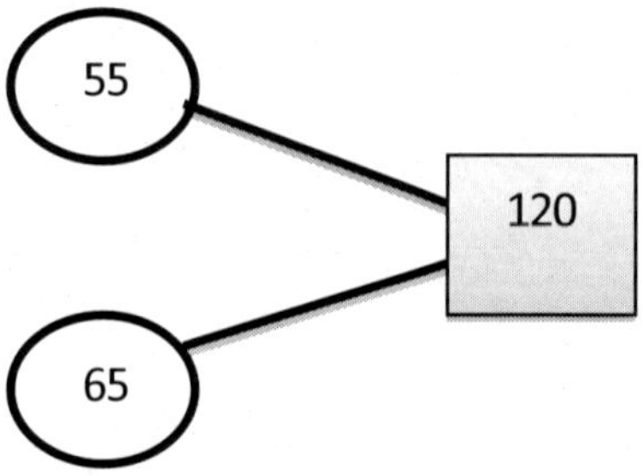

55 + 65 = 120

3. Circle the closest to 200.

228 (186) 2006 215 130

4. The number 6 goes in two of the

×	1	2	3	4	5
1					
2			X		
3		X			
4					
5					

squares of this multiplication grid. Cross (X) the two squares where 6 goes.

5. Complete these calculations.

20 x 10 = 200

200 x 10 = 2000

2000 ÷ 200 = 10

200 ÷ 20 = 10

6. Calculate

(12x8) ÷ 4 + 18 = 42

Use BODMAS

7. Amy buys two pens, each at the same price.

She pays £5 and gets £1.60 balance.

What is the cost of one pen?
£5 - £1.60 = £3.40
£3.4 / 2 = £1.70

8. Megan and Ben each have a bag.
Megan's bag weighs 2 ½ Kg.
Ben's bag weighs 2.3 Kg.
How many grams less does Ben's bag weigh than Megan's bag?
Answer - 2500g – 2300g = 200g

2 ½ x 1000 = 2500g
2.3 x 1000 = 2300g

9. Chen has a parcel.
20% of the parcel weighs 54 Kg.
What is the weight of the 50% of the parcel?
If 20% is equals to 54kg,
50% is = 54/20 x 50 = 135kg

10. Write these in the order of size, starting from the biggest.
2/4 0.35 3/5 0.65

0.65	3/5	2/4	0.35

3/5 = 0.6
2/4 = 0.5

11. What is the perimeter of the given shape?

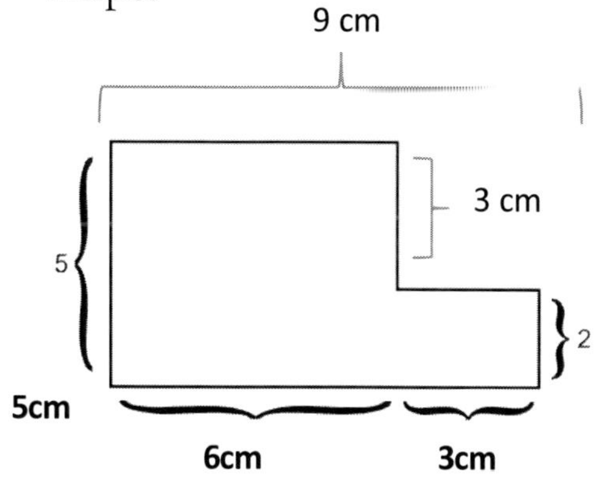

5+3+2+6+3+9 = 28 cm

12. Peter bought a pen, a pencil and a ruler for 85p.

The pen cost 15p more than the pencil.
The ruler cost 5p less than the pencil.

How much do a pen, a pencil and a ruler each cost?

Pen – 40p
Pencil – 25p
Ruler – 20p

If pencil = x
Pen = x+15
Ruler = x – 5
x+x+15+x-5 = 85
3x = 85 – 10
3x = 75
X = 25

13. What are the sizes of angels b and e?

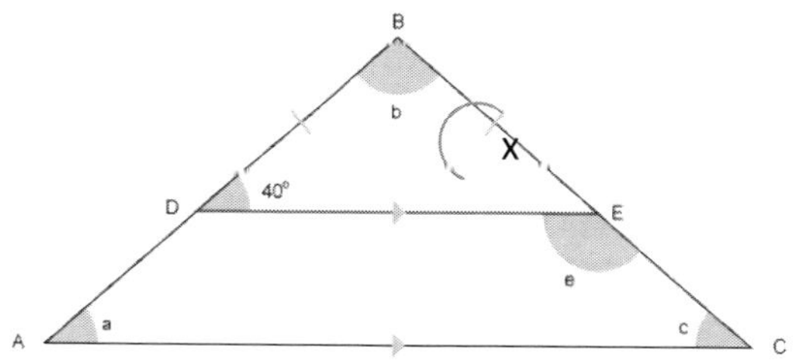

b- 100° e- 140°

Since BD and BE are equal x is = 40°

A triangle consist 180°

Therefore, b = 180 – (40+40)

= 180 – 80

= 100°

Likewise, a straight line is 180°

Therefore, e = 180 – x

= 180 – 40

= 140°

14. Sarah has a set of cards.

15 23 34 45

She chooses two of these cards and adds the number.

She rounds the result to the nearest 10.

Her answer is 70.

Which two cards did Sarah choose?

45 & 23

45 + 23 = 68

Nearest 10 is 70

15. Write these times in order, starting with the shortest.

15 weeks 3 days 52 hours 3 months

52 hours	3 days	3 months	15 weeks

52 hours is less than 3 days.
One day = 25 hours
Therefore, 52/24 = 2.2 days
15 weeks is more than 3 months
One month is equals to 4 weeks
Therefore, 15/4 = 3.75 months

16. Write the four missing numbers.

Note – to make the number in the square, multiply the two numbers in the connecting circles.

8*12 = 96

8*6 = 48

8*3 = 24

8*9 = 72

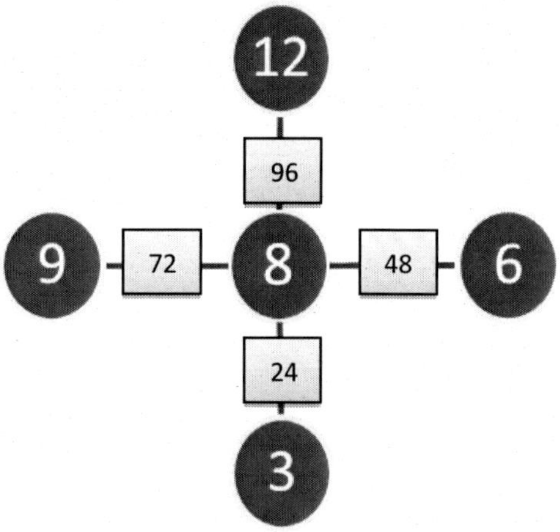

17. Draw the reflection of the shape in the mirror line.

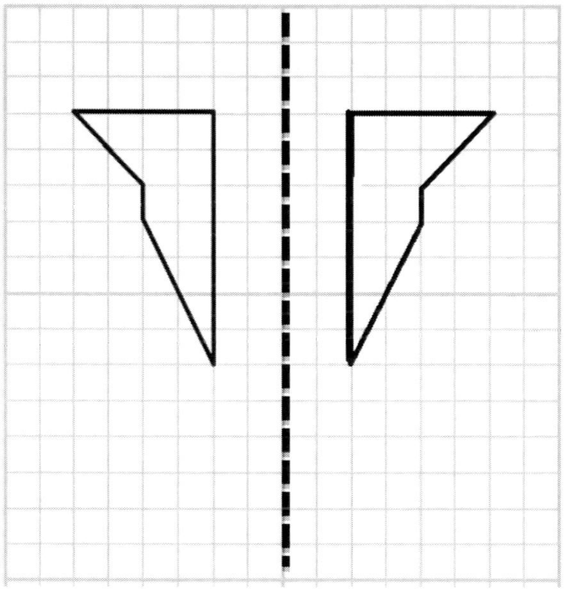

18. Provide the answer of the calculation.

9 + (5x2) ÷ 5 = 11
Use BODMAS

19. Martha buys a meal which includes the following.

Item	Cost
Burger	£4.15
French fries	£2.50
Coke	£1.95

What is the cost of the meal?

£4.15 + £2.50 + £1.95 = £8.60

20. Calculate.
925 ÷ 16 = 57.8125

21. The following table shows all the possible totals by adding all the possible combinations from Dice A and Dice B.

What is the probability of throwing a total 6?

Dice B \ Dice A	1	2	3	4	5	6
1	2	3	4	5	6	7
2	3	4	5	6	7	8
3	4	5	6	7	8	9
4	5	6	7	8	9	10
5	6	7	8	9	10	11
6	7	8	9	10	11	12

Provide the answer in percentage.

5 chances of getting 6 out of 36 possible outcomes.

Therefore,

5/36 = 0.139 *100 = 13.9%

22. The weight of potato sack is 11.25Kg.
What is the weight of the potato sack in grams? 11250 g

1000 grams = 1 kg

11.25 *1000= 11250 grams

23. $6y + 2 = 5y + 8$
What is the value of y? y= 6

$6y+2= 5y+8$

$6y-5y = 8-2$

$1y = 6$

24. Calculate.
$34.12 \div 1.2 = 28.43$ (2.d.p.)

Mathematics Paper 6

1. Christy has two boxes
There are X counters of chocolates in each box.

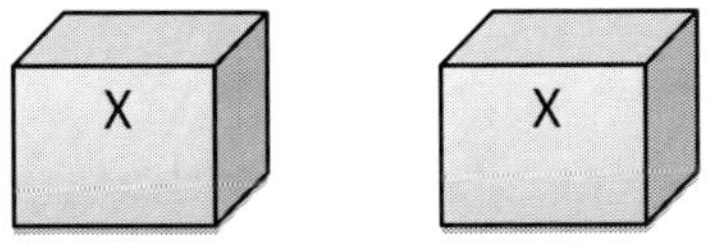

She puts all her chocolates in a bag, and then gives 3 chocolates to Seb.

Write an expression for the number of chocolates in the bag now.

2X - 3

2. Bill and Tom are brothers. Bill weighs two times more than Tom.

Together they weigh 96Kg.

How much does Bill weigh? 64kg

If Tom's weight is = x

Bill's weight = 2x

Therefore 3x = 96kg

X = 32kg

3. Here is a triangle on a square grid.
Draw the new position when it is rotated 270° about point X.

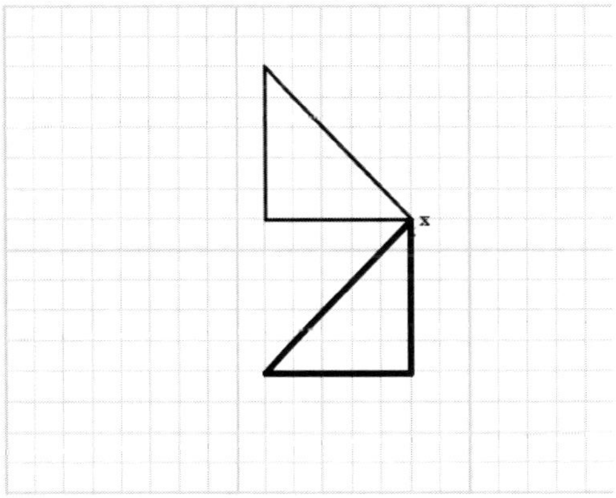

4. Amy has £120 to buy cupcakes and chocolates for her birthday.

Each cupcake costs £6.75
Each chocolate costs £3.75

She buys 15 cupcakes.
How many chocolates can she buy for the balance money? 4 chocolates
Total cost of cupcakes = 15 * £6.75 = £101.25
Balance money = £120 - £101.25 = £18.75
£18.75 / £3.75 = 5

5. Here is the graph information of three children.

a) Which child's information is missing in the graph?

Abby

b) Draw the graph with all four children's information.

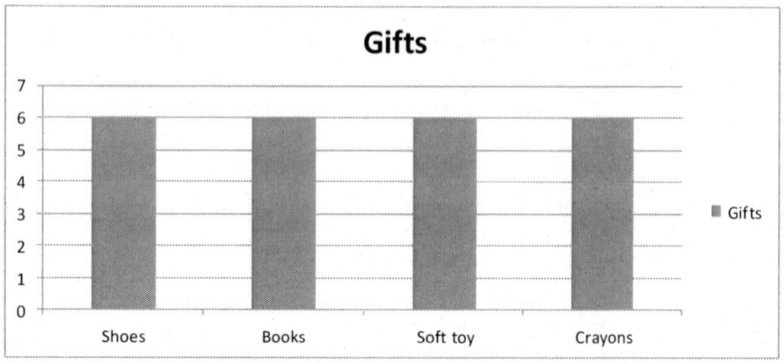

6. The diagram shows a square of side length 14cm.
Inside the square are 8 congruent trapeziums and a shaded square.

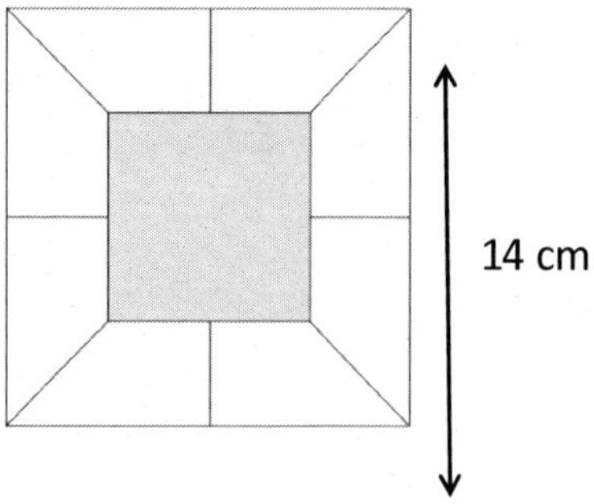

The area of a trapezium is 20cm2.
The height of a trapezium is 4 cm.
What is the area of the shaded square?

2 methods
Method A –
14 – 8 = 6cm

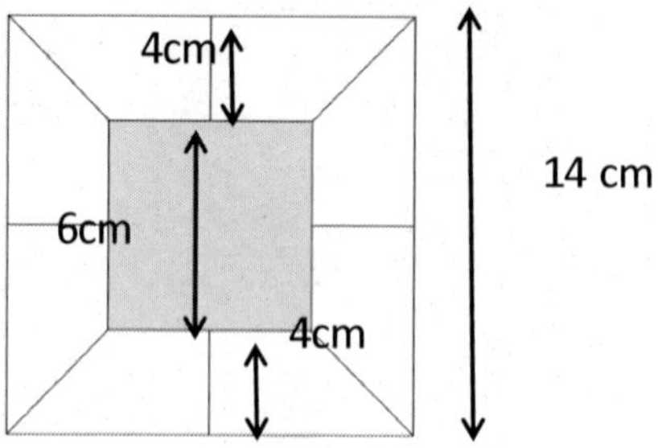

Therefore the area of the shaded square is = 6*6 = 36 cm2

Method B-
Area of the square is= 14*14= 196 cm2
Area of all trapezium is= 20*8 = 160cm2
Therefore, the area of the shaded square is = 196 -160 = 36cm2

7. Eighty children need 2 books and one pencil each.
There are 20 books in a box and 10 pencils in a bag.

How many boxes of books and bags of pencils are needed?

Boxes of books needed are= 80/20 = 4
Bags of pencils needed are = 80/10 = 8

8. The signs are missing from these number sentences.
Write the missing signs, **+, -, x and ÷.**

2nd column	6th column
+	X
+	X
X	+

9. Round the following numbers.
758 to the nearest 100 – 800
956 to the nearest 10 - 960
2 3/8 to the nearest whole number – 2
97.3962 to the nearest hundredth place –97.40

10. Lara goes to shop to buy peanuts.
Each bag of peanut cost £2.79.
How many bags of peanuts can Lara Buy for £10? 3 bags
10/2.79 = 3

11. Sara has a basket that contains 12 oranges and 14 apples.
What fraction of the fruits in the basket are apples?
14/26 = 7/13

Sara gives one apple and one orange to Rita.
What fraction of the fruits in the basket are oranges now?
11/24

12. Below is a recipe for cake.
– 250g butter, softened
– 1 cup caster sugar
– 2 teaspoons vanilla essence
– 3 eggs
– 2 1/2 cups self-raising flour
– 2/3 cup milk

Karen uses the recipe to make cake.
She uses 1 1/3 cups of milk.
How many eggs and vanilla essence will she use?
1 1/3 ÷ 2/3 = 2
Therefore, Eggs = 3 *2 = 6 eggs
Vanilla essence = 2*2 = 4 teaspoons

13. The below graph shows a survey conducted among a group of children in order to find out children's favourite fruit.

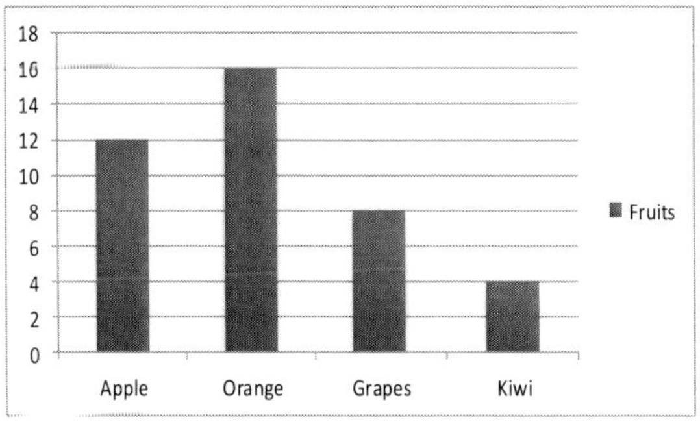

How many more children like oranges than grapes?

16-8=8

Rita says 'Exactly thrice as many children like oranges as children like kiwi'.

Is Rita correct? No

Explain how you know?

4 children like kiwi. 3 times four is = 12
But 16 children like oranges.
Therefore, Rita is wrong.

14. Sara takes a taxi to go to her aunt's place.

    It cost £1.50 for each kilometre.

    She pays the taxi £12.75.

    How many kilometres did she travel in the taxi? 8.5 km

    12.75/1.5 = 8.5

15. One double-digit number and two single-digit numbers multiply to make 576.

    Write the missing number?

    12 x 8 x 6 = 576

16. Rita and Sara have two 3D objects. Rita's object has 4 triangle faces. Sara's object has 4 triangle faces and one square.
    What are the names of the two objects?

    Rita – Tetrahedron
    Sara – Square-Based Pyramid

17. Write the missing number. 14

(23 +   ) x 28 = 1036

If the missing number is x
(23+x) x28 =1036
23+x= 1036 /28
23+x= 37
X = 37 – 23 = 14

18. Amy decides to build a pond in her garden.
    She orders 9000 bricks.
    The store can deliver 750 bricks in a truck at a time.

    How many times do they have to send the truck to Amy's place?

    9000/750= 12 times

    Amy returns 600 bricks. The cost of the balance bricks is £1260. How much does a single brick cost?

    9000- 600 = 8400 bricks

    8400 bricks = £1260

    Therefore, one brick is = 1260/8400 = £0.15

19. Emily has eight 5p coins and fifteen 10p coins.

How much money does she have altogether?

10x 15= 150

5x 8= 40

150+40 = 190 = £1.90

20. Sara multiplies two numbers and divides the result by three.
Her answer is 32.
One of Sara's numbers is 8.

What is Sara's other number? 12

If the other number is y,
(8*y)÷3 = 32
8*y– 32 *3 = 96
Y = 96/8 = 12

21.

Box of cereal   Box of Butter

£0.80   £0.40

Rita buys 6 boxes of cereal and 8 boxes of butter.

If she pays £10, how much balance will she get? £2

Total cost = (6*0.8) + (8*0.4)

4.8 + 3.2 = £8

Therefore, 10-8 = £2

22. Below shown is a grid with design.

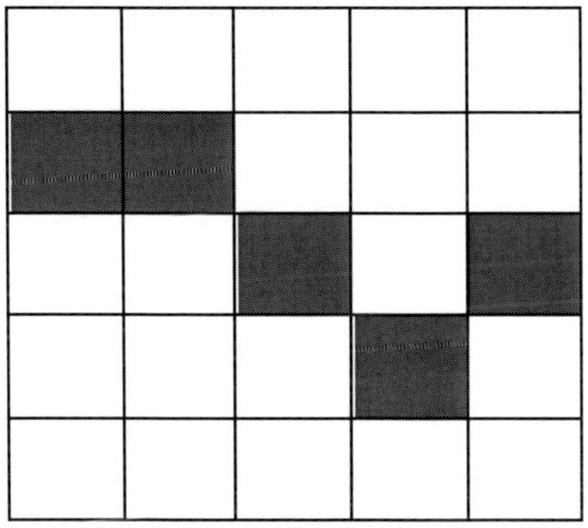

Sara rotates it to 90o. Shade in the missing parts of design.

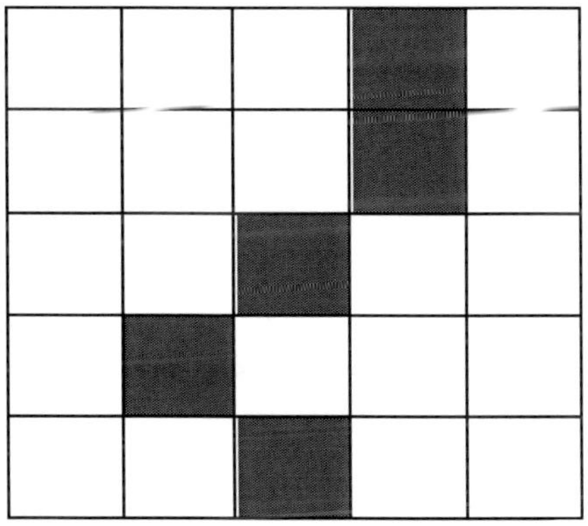

23. Rita has two equal size squares

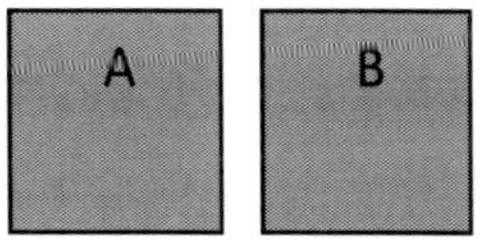

Each square's perimeter is 72cm.

She cuts square B in to half and attaches it with square A and makes a rectangle.

What is the perimeter of the rectangle?

72 + (72/2) = 108 cm

24. Cross (x) the shape that is exactly ¼ shaded

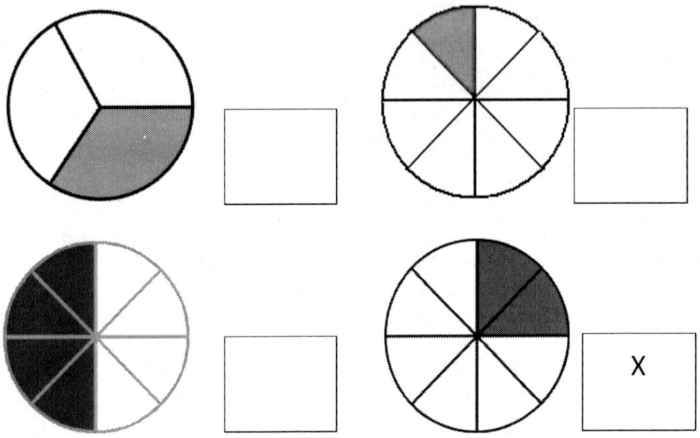

Mathematics Paper 7

1.

| £4.50 | £4.55 | £45 | £45.50 |

1 mark

2. 2a south China tigers

2b Indochinese tigers

Up to 2 mark

3. 372

1 mark

4.

($2\frac{1}{2}$ + $4\frac{1}{2}$) * 7

= 49

1 mark

5. 46 is 4 less than 50, but 56 is 6 more

56 is nearer to 60, but 46 is nearer to 50

1 mark

6. 3843

1 mark

7.

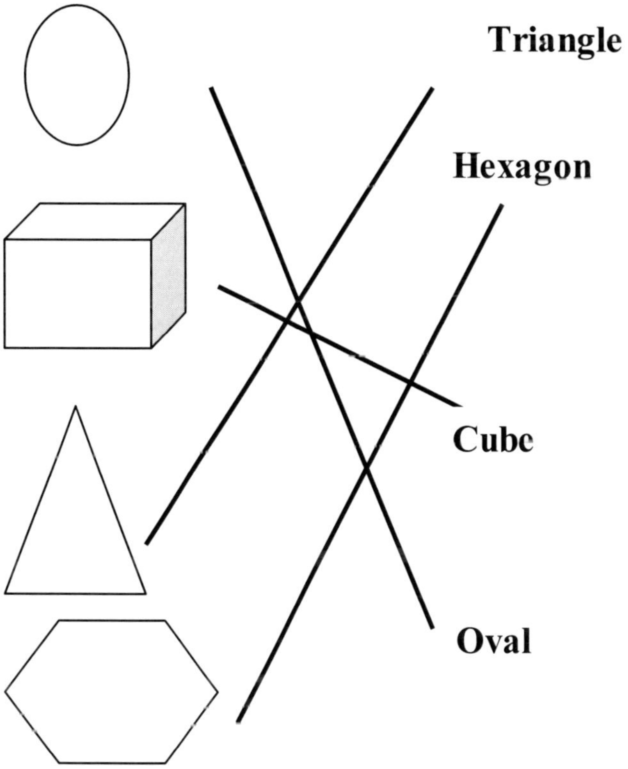

2 marks

8. 50, -100

1 mark

9. 9/11 3/6 2/3 1/9 16/20

2/3

1 mark

10. 100

1 mark

11. 101

1 mark

12.

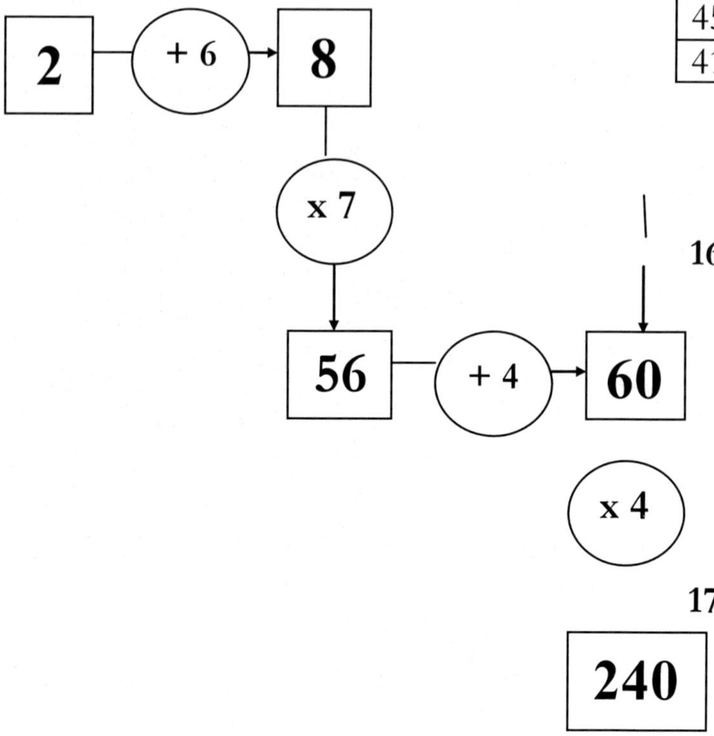

2 marks

13. 2x9 + 2a = 60

18 + 2a = 60

2a = 60 – 18

2a = 42

a= 21

2 marks

14. 112

1 mark

15.

	Rounded to the nearest 100
475	500
4955	5000
45565	45600
415289	415300

2 marks

16.

£ 49.75 + £ 37.50 + £ 12.85 = £ 100.10

1 mark

17. a £ 1.50

1 mark

b £ 16

2 marks

18.

2 x Dolls + £ 17.50 = £ 50

2 x Dolls = £ 50 - £ 17.50

2 x Dolls = £ 32.50

1 Doll = £ 32.50/2

1 Doll = £ 16.25

2 marks

19.

144 121 100 [81] [64]

2 marks

20.

a .25

1 mark

b.

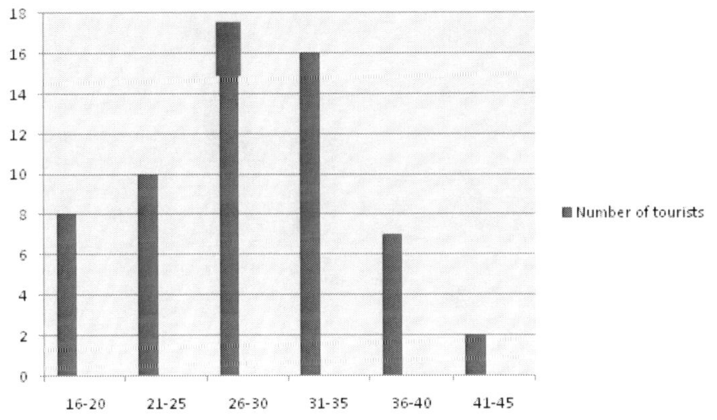

1 mark

21.

£5 – (£2.75 + £1.80) = x

£5 - £4.55 = x

x = £0.45

2 marks

22. 60

1 mark

23.

Yes **No** (circled)

No mark is awarded for circling 'No' alone.

An explanation which gives a counter-example to illustrate that not all numbers ending in 6 are multiples of 6, eg:

16 is not a multiple of 6'

2 marks

24.

(0.15) 0.35 0.50 0 (0.65) 5

1 mark

25.

X – 2975/35

X = 85

Mathematics Paper 8

1.

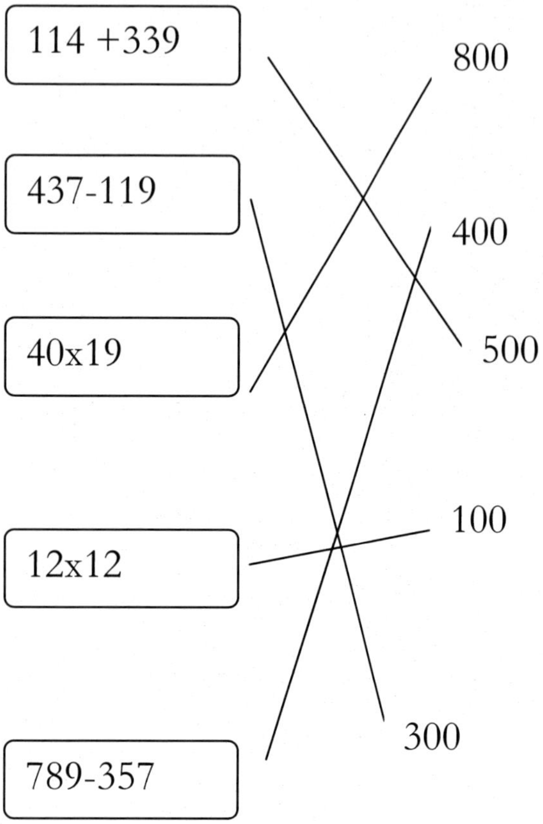

2 marks

2.

a 16

b

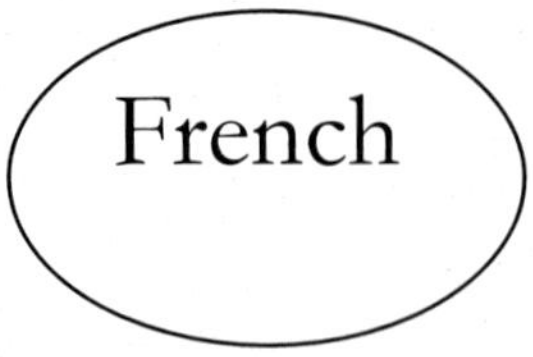

Up to 2 mark

3. 4.5

800

1 mark

4.

a40015cm

b. 90°

Up to 2 mark

5. £97.75

1 mark

6. 18

1 mark

7. £214.50

1 mark

8.

10 + (10+5) + (15+5) + (20+5)

=70

2 marks

9.

17	X 30 = 510

1 mark

10.

2	10	34	106	322

2 marks

11.

Area = (20cmx9)/2

=90cm^2

1 mark

12.

a. X = 180 – 90 – 49
   X = 41

b. Y = 180 – 41 = 139

2 marks

13.

Any numbers above 100 and can be the sequence of both is eligible for marks,

108, 120, 132, 144 ……

2 marks

14.

(£5.25 + £3.65 + £7.50 + £10.75 + £9.20)/5 = £7.27

2 marks

15.

X = 5Y,

X - Y = 360

We can replace X with 5Y

5Y – Y = 360

4Y = 360

Y = 90

a.
= (5x90) = 450

b.
= 90

2 marks

16.

160000/5000

=32

2 marks

17.

10 hours

1 mark

18.

$\text{Area}_x = (3\text{ cm})(5\text{ cm}) = 15\text{ cm}^2$

$\text{Area}_y = (3\text{ cm})(8\text{ cm}) = 24\text{ cm}^2$

$\text{Area}_z = (5\text{ cm})(8\text{ cm}) = 40\text{ cm}^2$

Total Surface Area = 2(15 cm^2) + 2(24 cm^2) + 2(40 cm^2)

Total Surface Area = 30 cm^2 + 48 cm^2 + 80 cm^2

Total Surface Area = 158 cm^2

2 marks

19.

X = 500 cows – 60 cows +18 sheep's

X = 440 cows + 18 sheep's

X = 458 animals

2 marks

20.

a. 3
b. Ann

2 marks

21.

4k +6(j −2)

= 4x3 + 6(5-2)

= 12 + 6x3

= 12 + 18

= 30

2 marks

22. 211.81

1 mark

23.

1 torch + 3 batteries = £ 16.90

1 torch = £ 10.00

So

£ 10.00 +3 batteries = £ 16.90

3 batteries = £ 16.90 - £ 10.00

3 batteries = £6.90

1 battery = £6.90/3

=£2.30

2 marks

Mathematics Paper 9

1. 4.78

1 mark

2. (64 + 36)= 100

1 mark

3. 24cm^2

1 mark

4.

Longest

2 ½ months → 6 weeks → 26 days → 36 hours

1 mark

5.

X = 50000 + 1000x7

X = 57000L

Up to 2 marks

6.

1 hour 10 minutes + 2 hour 7 minutes

= 3 hour & 17 minutes

1 mark

7.

a. How many faces it has ? 3

b. How many edges it has? 2

2 marks

8.

X = 100 – (13+35+27)

X = 100 – 75

X = 25

2 marks

9. Write the next two numbers in Sarah's sequence.

64 58 52 **, 46, 40**

2 marks

10.

(17/20) 17/25 19/20 19/25

1 mark

11.

100/9

=99

So 100-99

=1

1 mark

12. 123x69 = 8487

2 marks

13. a. 6% b. 67%

2 marks

14.

6 cm^2 = (2cm x base)/2

Base = 12 cm^2/2cm

Base = 6cm

2 marks

15.

3, 3/4 + 2, ½

= 15/4 + 5/2

Since the denominators are 4 & 2. We need to convert in to equal denominators. For this reason; let's multiply 5/2 by 2/2. This will give us 10/4. Using this we can continue our calculation,

15/4 + 10/4

=25/4

=6 ¼

2 marks

16.

Any line meeting X & which will exceed 180 angle are eligible for 2 marks.

2 marks

17.

Cumulative sales = £510 + £630 + £370 + £540 + £810 + £1230

Cumulative sales = £4090

2 marks

18.

£112.85 + £19.25

=£132.10

1 mark

19.

£1.20/24

=£0.05

1 mark

20.

1200g/2 = 600g

So

600g – 350g

=250g, which is the weight of Box B

2 marks

21. 5g

1 mark

22.

The difference between the measures will give the length of the small square,

54cm – 42cm

= 12cm

So the length of the large square would be,

42cm – 12cm

= 30cm

2 marks

23.

Sarah's book expense = £10 - £2.65 change = £7.35

Lisa's book expense = £12.50

So the difference between the costs is,

£12.50 - £7.35

=£5.15

2 marks

24. 288.50

1 mark

25. B, C, A, D

1 mark

26.

An explanation which recognizes that 20% of the number must be multiplied by 5 or equivalent,

eg:

'You multiply by 5'

Mathematics Paper 10

1.

1500 ÷ 5 =	✓
5500 ÷ 11 =	✓
750 ÷ 8 =	
225 ÷ 3 =	
18500 ÷ 175 =	✓

Up to two marks

2. 7584

1 mark

3. a. Ascending order

3/10, 3/9, 3/7 , 3/4

b. Descending order

3/4, 3/7, 3/9, 3/10

Up to two marks

4. 4050kg

1 mark

5.

a. 6+4 = 10
b. 2+5 = 7

1 mark

6. 26 and 65

Up to two marks

7.

a. 150/13.79
=10.87

So the answer is 10 bags of coffee

b. 1 coffee machine = £68.95
3 bags of coffee = £13.79 x 3
2 bags of sugar = £5.35 x 2

So the expenditure of all these items equals to,
£68.95 + £41.37 + £10.70 = £121.02

Sam has £50 but the total cost is £121.02

So the additional cost would be,

£121.02 - £50
=£71.02

Up to two marks

c. 0.60

1 mark

d. 67.50 km

1 mark

e.

$a^2 = b^2 + c^2$,

b = 7 and c = 20

So

$a^2 = 7^2 + 24^2$

$a^2 = 49+576$

$a^2 = 625$

a = 25

Up to two marks

f. 14-8 = 6

Circle Yes (circled) No

Children who liked fiction were 14. On the other hand 7 children liked nonfiction. This indicates us twice as many children chose fiction as chose nonfiction.

Up to two marks

g. 12.17 am

1 mark

h.

One has been done for you.

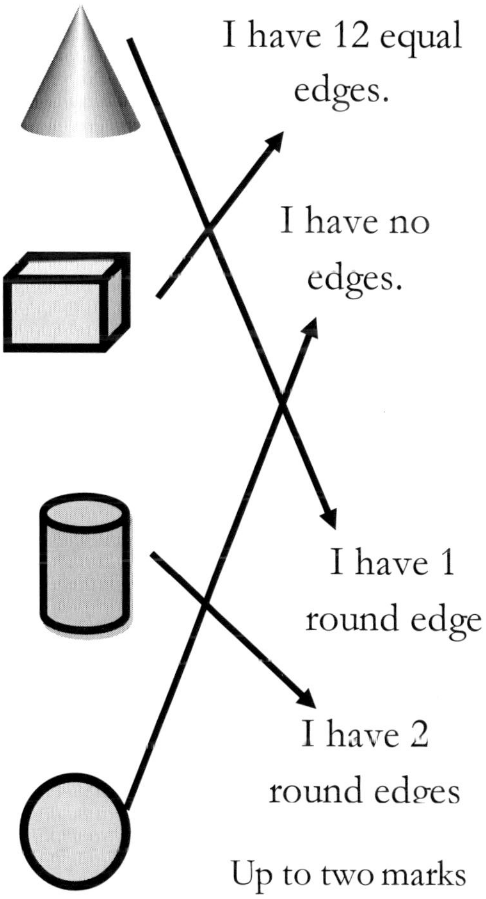

Up to two marks

i. X = 0.6

1 mark

17 x 27 = 459

Up to two marks

j.

We need to apply the BODMAS principle here

So the calculation needs to be done in the following way,

First divide, then multiply and finally add.

5 + 16 ÷ 4 × 2

= 2.5 + 4 x 2

= 2.5 +8

=10.5

Up to two marks

k. ABC

1 mark

8. 2760

1 mark

9.

6 x 6 = 36

7 x 7 = 49

8 x 8 = 64

Here the 49 is closer to 50, so the square of 50 will be closer to 7.

The correct answer is 7.1, which is giving a close answer of 50.41.

2 marks

10.

C = 2πr
C = 2 x 22/7 x 14

C = 88 m

2 marks

11.

a. 4
b. (3 x 60 + 5 x 24)/8
= (180 + 120)
= 300/8
= 37.50mph

Up to two marks

12. 8/10

2 marks

13.

X	1	2	3	4	5
1					
2					
3			✓		
4					
5					

2 marks

Mathematics Paper 11

1. £3.50, £5.30, £30.50, £32.50 (1 mark)

2. 268 (1 mark)

3. 263 (1 mark)

4. 90 - (9 + 20 + 18) = 43 (2 marks)

5. 20 x 0.20 = £4.00 (2 marks)

6.

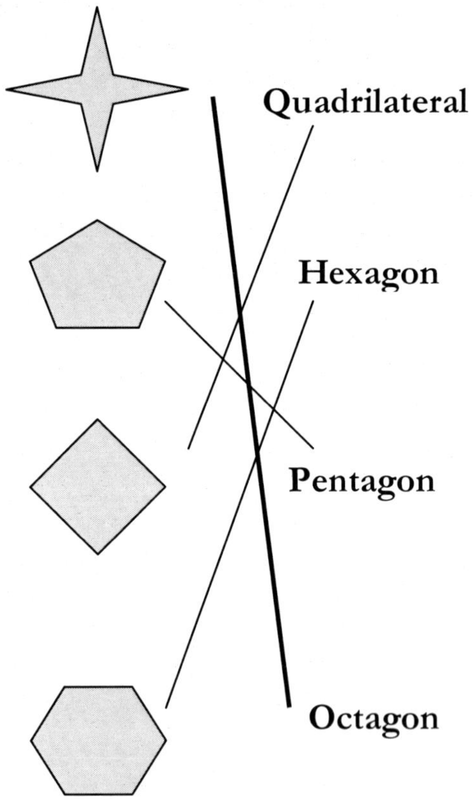

(2 marks)

Quadrilateral = 4 sided shape
Hexagon = 6 sided shape
Pentagon = 5 sided shape
Octagon = 8 sided shape

7. 496 (1 mark)

8. A = 5
B = 7
C = 25
D = 50 (2 marks)

The first step in this question is 21 divided by 3 to get answer B. Then take away 2 to get answer A. Then 21 add 4 to get answer C and finally multiply answer C by 2 to get answer D.

9. 08:45 (1 mark)

10. 0.53 and 0.39 (2 marks)

11. 112 (2 marks)

Simpler calculation is 2 x 280 divide by 5.

12. 122 (1 mark)

13. a). 24/6 = 4 boxes (1 mark)

b). 20 (1 mark)

1/6th of 24 = 4, then 24 – 4 = 20

14. £12.75 (2 marks)

(2 x 6.50) + 4.25= 13 + 4.25 = 17.25

30 – 17.25 = 12.25

15.

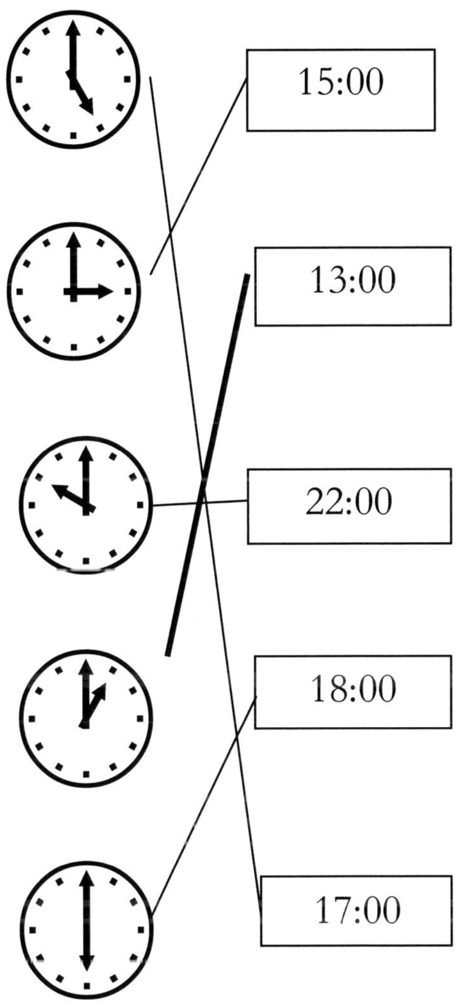

(2 marks)

16. a). 80p (1 mark)

£7.20 divide by 9 equals 80p

b). £4.80 (2 marks)

1/3 of 7.20 = 2.40

7.20 – 2.40 = 4.80

17. 32, 16, 36, 56, 28 (2 marks)

18. 12, 9 (1 mark)

19. 6 months, 2 weeks, 72 hours, 2 days (2 marks)

20. 690 (1 mark)

21. 124, 112 (1 mark)

136 – 12 = 124 – 12 = 112

22. 101 (1 mark)

1212 divide by 12

23. a). 95p (2 marks)

£1.75 + 80p = £2.55

£3.50 – £2.55 = 95p

b). £8.40 (2 marks)

(£1.75 x 2) + £4.50 + (2 x £0.20) =

£3.50 + £4.50 + £0.40

24. 48 (1 mark)

(0.4 x 120)

Mathematics Paper 12

1. a). 1000 (1 mark)

b). 50.60 (1 mark)

2. a). 45+12+9+4=70 (1 mark)

b). CAT (1 mark)

3. 0.6 (1 mark)

4. Liam = 1.49 m

Visha = 1.36 m

Jack = 150 cm (2 marks)

5.

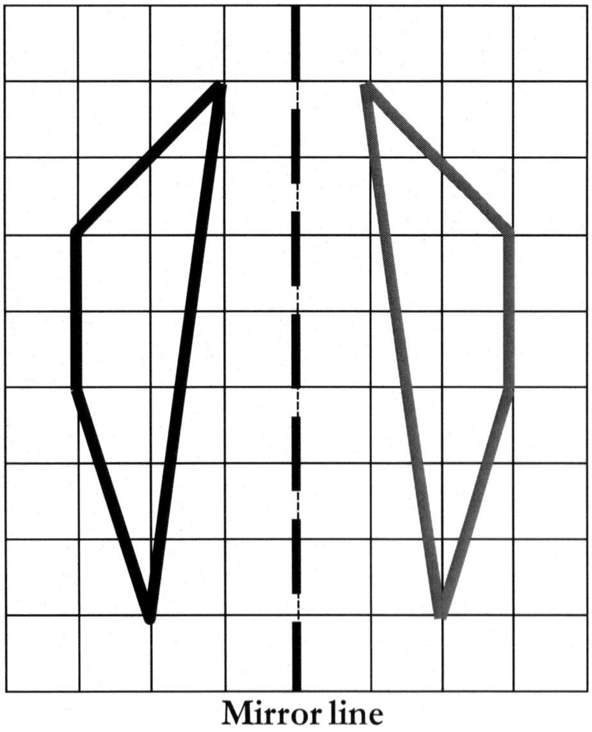

(2 marks)

6. 301.50 (2 marks)

£670 x 0.45

7. 3 pm in Canada (2 marks)

If the flight left at 11 am, you then add 9 hours to 11am to give you the time of landing in London time, which is 8pm. The question states that the time in Canada is 5 hours in front of London time, so you then take away 5 hours from 8 pm to give you 3 pm in Canadian time.

8. 555 (1 mark)

525 + 30

9. 2, 51 (2 marks)

10. (12, 3) (2 marks)

11. X = 35° Y = 145° (2 marks)

X = 90 – 55

Y = 180 – 35

Note to parent: With this question it is important that your child understands that the angle about a straight horizontal line from any point is 180° so in this example 35° + 145° = 180°

12. C (1 mark)

13. a). 74 (1 mark)

b). 150 (1 mark)

14. 82 (1 mark)

15. S = 240

T = 80 (2 marks)

S = 3T, S – T = 160

3T – T = 160 so T = 80.
Therefore 3T = 240

16. 10 (2 marks)

17. 9 (1 mark)

18. 08:50 (2 marks)

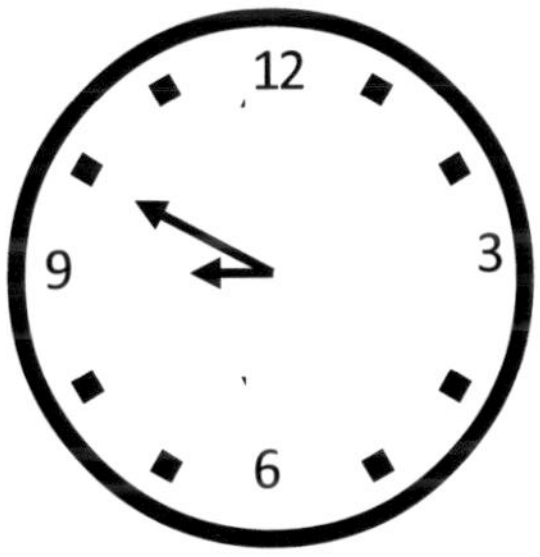

19. 5 (1 mark)

20. 214 (2 marks)

250 – 50 = 200 then 200 + 14

21. £10 (1 mark)

£2.50 x 4

22. 378 (2 marks)

23. 10th January (1 mark)

From the 3rd January add 7 days

24. 26 slices (2 marks)

13 people get 2 slices each therefore
13 x 2

Mathematics Paper 13

1. £1.92 (1 mark)

£1.20 divide by 5 then multiply by 8.

2. 0.11, 0.26, 0.52, 0.82 (1 mark)

3. 36 years old (2 marks)

78 take away 6 then divide by 2.

4.

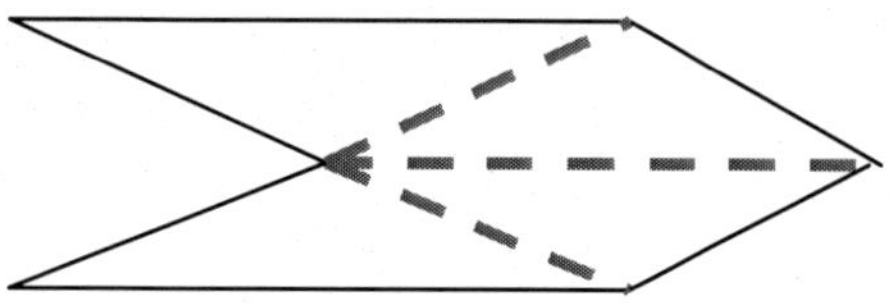

(1 mark)

5.

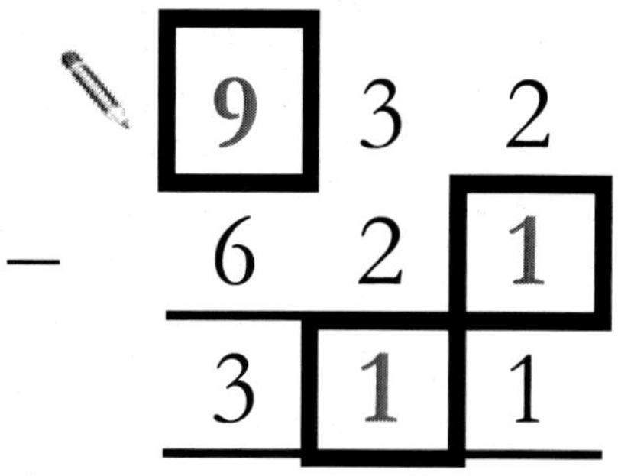

(2 marks)

6. a). £40 original price (2 marks)

A x 0.6 =24, therefore 24 divide by 0.6 gives 40.

b). £16 (1 mark)

40 take away 24.

7. a).

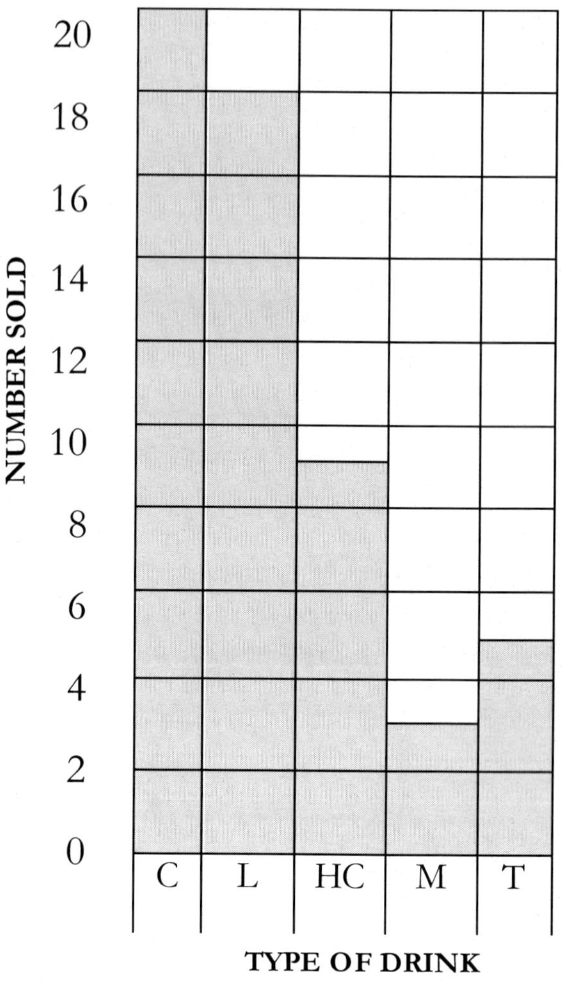

(2 marks)

b). 32 cups (1 mark)

20 + 9 +3

8. 727 (1 mark)

9. 206 (1 mark)

10. Muffin = 40p and Cookie = 25p

(2 marks)

11.

Number	Rounded to the nearest **whole** number
6.05	6
6.55	7
5.45	5
5.54	6

(2 marks)

12. £30 (2 marks)

(23 + 37) divide by 2.

13. £18.95 (2 marks)

£10.50 + £5.25 + £3.20

14.

	2	7	2
+	4	5	6
	7	2	8

(1 mark)

15. 11 (1 mark)

16. 800 (1 mark)

Find the place you want to round. Look one place to the right. If the digit is less than 5, round down. If the digit is 5 or higher, round up. First, find the digit in the hundredth place. This is the place you want to round.

17. 6000 (1 mark)

2000 + 4000

18. a). £29.70 (1 mark)

9 multiply by £9.90.

b). £29.40 (1 mark)

6 multiply by £4.90.

c). £59.10 (1 mark)

£29.70 + £29.40

d). 8 people (1 mark)

£4.80 divide by £0.60

19. 32 (2 marks)

(6 x 6) =36, (2x2)= 4, then 36 – 4 = 32.

20. £4.85 (2 marks)

£9.50 - £2.25 – (£1.20 x2)

21. 240 miles (1 mark)

60 multiply by 4

22. 1000 m (1 mark)

1 hour = 60 minutes
The car travels 60 km in 60 minutes, so

in one minute the car would have travelled 1 km. 1 km = 1000m

23. £24.60 (2 marks)

0.2 x 20.50 = £4.10.

Therefore £4.10 + £20.50

24. 32 (2 marks)

James gave ¼ of the stamps given to him to Peter and kept the remaining 6. James kept 6 which is ¾ and Peter got ¼ which is 2. James initial number of stamps is 6 add 2 which is 8. Therefore James got 8 from John. John gave ½ of his stamps to James. We now know that James has got 8. So John initially had 2 multiply by 8 which is 16. Andy gave ½ of his stamps to John. We know that John got 16, so Andy's initial number of stamps is 2 multiplied by 16, which is 32.

Mathematics Paper 14

1. Elephant weighs 525 kg (2 marks)

700 divide by 4 as it's a 3:1 ratio, which equals 175 kg. Therefore the weight of the crocodile is 175 kg. Multiply 175 by 3 to get the elephants weight.

2.

Mirror line

(2 marks)

3. 336 (1 mark)

4. 4.6 (1 mark)

The difference between them is 0.1.

5. a). 130 ice creams (1 mark)

b). 50 more (1 mark)

6. £1.50 (2 marks)

14 x 0.5 = £7. Then £7 £5.50.

7. a). 3 teams (1 mark)

b). St Catherine's School (1 mark)

King Henry's School scored
2 + (3 x 1) = 5

Timmins scored
(2 x 2) + (2 x 1) = 6

St Catherine's School scored
(3 x 2) + (3 x 1) = 9

Harpur School scored
(4 x 2) = 8

Newham School scored
(2 x 1) + (4 x 1) = 6

8. £1.81, £2.24, £2.66, £2.69, £2.81

(2 marks)

9. $8.7 < 9.7$ 'less than' (1 mark)

10. 1.5 kg (1 mark)

5 multiply by 0.3.

11. £10 (2 marks)

For 25% off:

0.25 x 15 = 3.75
15 – 3.75 = 11.25

For 1/3 off:

1/3 of 15 is 5 ie 15 divide by 3.
15 – 5 = 10

12. 398 (1 mark)

X	1	2	3	4	5
1					
2					
3				✓	
4			✓		
5					

13. (2 marks)

14. 1284 (1 mark)

15.

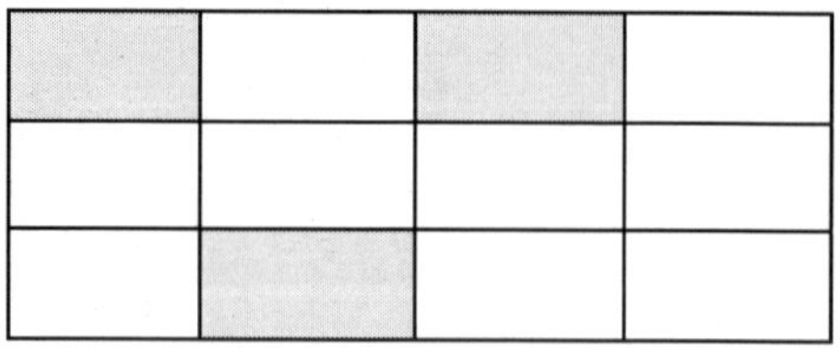

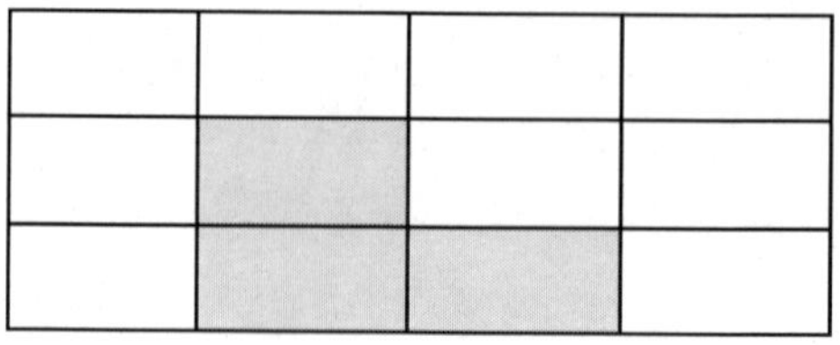

(2 marks)

There are 12 rectangles in each box. A quarter of 12 is 3. Therefore each box with 3 rectangles shaded is equal to a ¼ shaded.

16. £25 (2 marks)

Half of £5 is £2.50. Then multiply £2.50 by 10

17. 5p (2 marks)

£1.50 divide by 6 to give 25p per can in a 6 pack. Individually the coke cans are sold at 30p each. There the saving would be 30p – 25p.

18. £1600 (1 mark)

£2050 – £450

19. (2 marks)

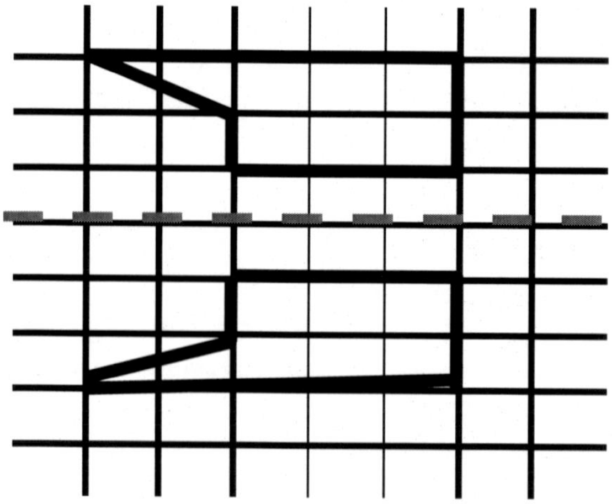

20. 0.6 (2 marks)

15 squares coloured : 25 squares in total

So a ratio of 3 : 5 equal to a fraction of 3/5 which equals 0.6.

21. £200 (2 marks)

Monday : 3 x 10 = 30
Tuesday : 4 x 10 = 40
Wednesday 5 x 10 = 50
Friday 8 x 10 = 80

Therefore 30 + 40 + 50 + 80

22. 18 – 6 = 2 (2 marks)

23. 1 (1 mark)

24. 2 hours 55 minutes (2 marks)

Mathematics Paper 15

1. 33 (1 mark)

2. a). 15 cars (1 mark)

5 multiplied by 3

b). 85 cars (2 marks)

17 x 5 = 85

c). 20 % (2 marks)

5 out of 25. Therefore divide 2 by 25 then multiply by 100.

3. 5p, 25p, £1.20, £20, £150 (1 mark)

4. 3745 (1 mark)

5. 62 (1 mark)

6. 3 and 5 (2 marks)

7. a). 6 b). 52 c). 48 (2 marks)

8. 294 (2 marks)

14 multiplied by 21

9. a). (2 marks)

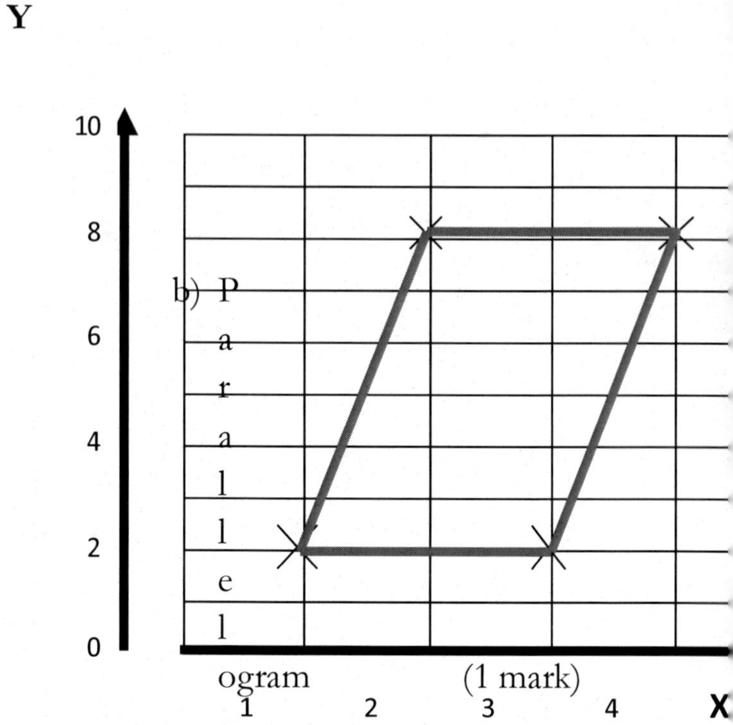

b) Parallelogram (1 mark)

10. a). 1:10 (10 minutes past 1 o'clock)

(1 mark)

b). 09:55 (55 minutes past 9) (1 mark)

11. Multiples of 4 is 12, Multiples of 5 is 25, Multiple of both 4 and 5 is 20, Multiple of neither is 53.

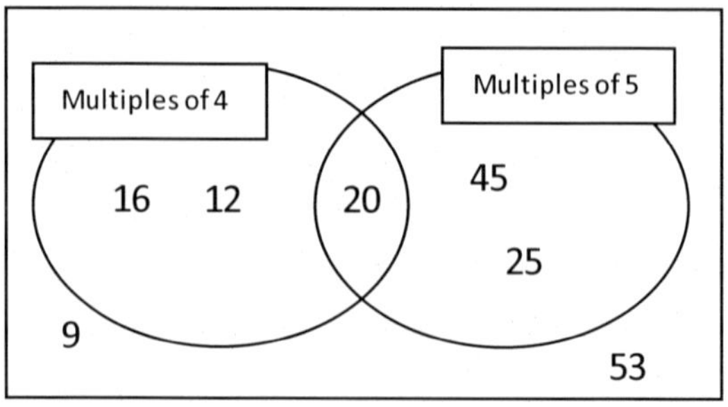

(2 marks)

12. 1284 (1 mark)

13. 11.45 (1 mark)

14. Hundreds place should be ticked. (1 mark)

15. 0.4, 5.4, 64.5 (1 mark)

16. ½ (half) (1 mark)

17. C (1 mark)

18. 1420g (1 mark)
(350 + 600 + 470)

19. 7.2 m (2 marks)
20 x 5 = 100 cm
1m = 100 cm
6.2 + 1 = 7.2 m

20. 900g (2 marks)
1,500 divide by 5 then multiply by 3

21. 13 (2 marks)

22. X= 4 (1 mark)

23. 219 and 381 (2 marks)

24. £3.50 (1 mark)
30 take away 26.50

25. 10,000 (1 mark)
2000 + 4000 + 6000

Mathematics Paper 16

1. 35 (1 mark)

2. 537 (1 mark)

3. $\frac{1}{2}$ $\frac{8}{16}$ $\frac{4}{8}$ $\frac{12}{24}$ (2 marks)

4. a). £2.20 (1 mark)

£7.80 – £5.60

b). £6.50 (1 mark)
£6.50 – £5.20

5.

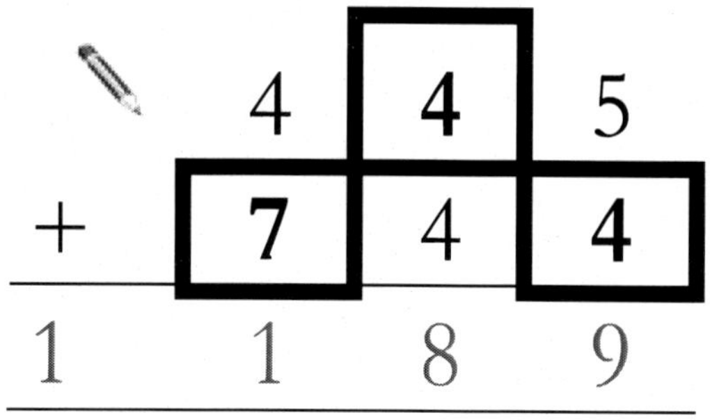

(2 marks)

6. 26 and 36 (2 marks)

7. 4080, 2590 and 840 (2 marks)

8. a). >, b). <, c). = (2 marks)

9. 200 (1 mark)

420 divide by 2.1

10. 3/4 (2 marks)

8 is unshaded out of 12 therefore divide both by 4 to get the lowest fraction of 2/3

11. £125 (2 marks)

0.25 x 500

12. 1.89 (1 mark)

13. X = 7 (2 marks)

32 + 10 then divide by 6

14.

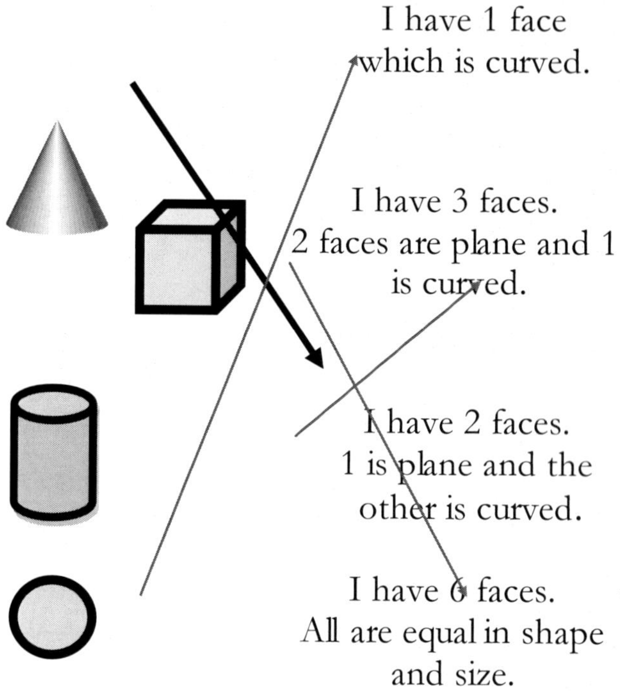

(2 marks)

15. £172.50 (2 marks)

(30 x 5) + (3 x 7.50) = £150 + £22.50

16. 75 ml (2 marks)

3 x 150 then divide by 6

17. a). 220, b). 100, c). 8, d). 40

(2 marks)

18. £4.50 (1 mark)

40.50 divide by 9

19. 29, 37, 45 (2 marks)

20. 13:50 (2 marks)

(10 minutes to 2 o'clock)

21. 662 (1 mark)

22. 18 (1 mark)

0.05 x 360

23. 241 (1 mark)

24. Multiples of 3 is 21.

Multiples of 8 is 48.

Multiples of both 3 and 8 is 24.

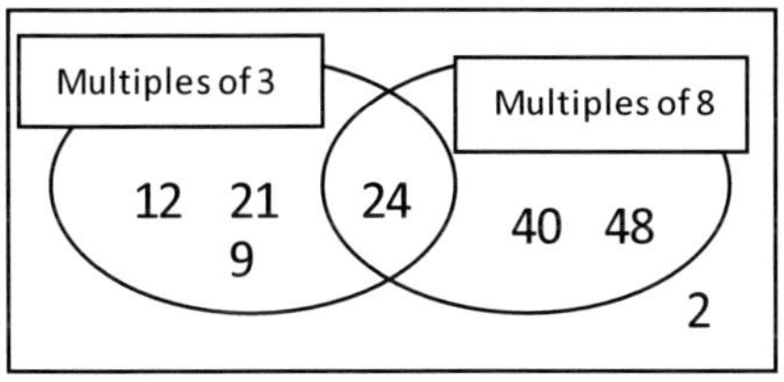

(2 marks)

Made in United States
North Haven, CT
05 November 2026

21772482R00174